"Race" and Early Childhood Education

Critical Cultural Studies of Childhood

Series Editors:
Marianne N. Bloch, Gaile Sloan Cannella, and Beth Blue Swadener

This series will focus on reframings of theory, research, policy, and pedagogies in childhood. A critical cultural study of childhood is one that offers a "prism" of possibilities for writing about power and its relationship to the cultural constructions of childhood, family, and education in broad societal, local, and global contexts. Books in the series will open up new spaces for dialogue and reconceptualization based on critical theoretical and methodological framings, including critical pedagogy, advocacy and social justice perspectives, cultural, historical and comparative studies of childhood, post-structural, postcolonial, and/or feminist studies of childhood, family, and education. The intent of the series is to examine the relations between power, language, and what is taken as normal/abnormal, good and natural, to understand the construction of the "other," difference and inclusions/exclusions that are embedded in current notions of childhood, family, educational reforms, policies, and the practices of schooling. *Critical Cultural Studies of Childhood* will open up dialogue about new possibilities for action and research.

Single authored as well as edited volumes focusing on critical studies of childhood from a variety of disciplinary and theoretical perspectives are included in the series. A particular focus is in a reimagining as well as critical reflection on policy and practice in early childhood, primary, and elementary education. It is the series intent to open up new spaces for reconceptualizing theories and traditions of research, policies, cultural reasonings and practices at all of these levels, in the Unites States, as well as comparatively.

Published by Palgrave:

The Child in the World/The World in the Child: Education and the Configuration of a Universal, Modern, and Globalized Childhood
Edited by Marianne N. Bloch, Devorah Kennedy, Theodora Lightfoot, and Dar Weyenberg; Foreword by Thomas S. Popkewitz

Beyond Pedagogies of Exclusion in Diverse Childhood Contexts: Transnational Challenges
Edited by Soula Mitakidou, Evangelia Tressou, Beth Blue Swadener, and Carl A. Grant

"Race" and Early Childhood Education: An International Approach to Identity, Politics, and Pedagogy
Edited by Glenda Mac Naughton and Karina Davis

"Race" and Early Childhood Education

An International Approach to Identity, Politics, and Pedagogy

Edited by

Glenda Mac Naughton and Karina Davis

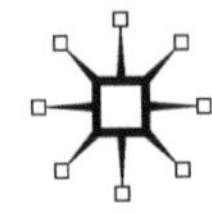

"RACE" AND EARLY CHILDHOOD EDUCATION

First published in 2009 by
PALGRAVE MACMILLAN®
in the United States—a division of St. Martin's Press LLC,
175 Fifth Avenue, New York, NY 10010.

Where this book is distributed in the UK, Europe and the rest of the world, this is by Palgrave Macmillan, a division of Macmillan Publishers Limited, registered in England, company number 785998, of Houndmills, Basingstoke, Hampshire RG21 6XS.

Palgrave Macmillan is the global academic imprint of the above companies and has companies and representatives throughout the world.

Palgrave® and Macmillan® are registered trademarks in the United States, the United Kingdom, Europe and other countries.

ISBN: 978–0–230–61324–9

Library of Congress Cataloging-in-Publication Data

"Race" and early childhood education : an international approach to identity, politics, and pedagogy / edited by Glenda Mac Naughton and Karina Davis.
p. cm.—(Critical cultural studies of childhood)
Includes bibliographical references and index.
ISBN 0–230–61324–1
1. Racism in education. 2. Early childhood education—Social aspects. 3. Race awareness. I. Naughton, Glenda Mac. II. Davis, Karina. III. Series.

LC212.5.R33 2009
372.1829—dc22 2008055773

A catalogue record of the book is available from the British Library.

This book is printed on paper suitable for recycling and made from fully managed and sustained forest sources. Logging, pulping and manufacturing processes are expected to conform to the environmental regulations of the country of origin.

Design by Newgen Imaging Systems (P) Ltd., Chennai, India.

First edition: August 2009

Transferred to Digital Printing in 2012

Glenda wishes to dedicate this book to Louise Derman-Sparks, Babette Brown, and Elizabeth Dau whose determined antiracist work has inspired, challenged, and affirmed my work for more racially just meaning-making and actions in the early childhood field over many years.

Karina wishes to dedicate this book to Kiana and Tahli whose experiences, desires, conversations, and questions are what help to confirm and reconfirm to me the importance of working toward greater racial justice for all of us.

Contents

Series Editors' Preface ix

Acknowledgments xi

Contributors xiii

Introduction: Thinking Differently: The Call and the Desire 1
Karina Davis and Glenda Mac Naughton

Part I Exploring the Politics of Children's Racialized Identities

1 Discourses of "Race" in Early Childhood: From Cognition to Power 17
Glenda Mac Naughton and Karina Davis

2 Exploring "Race-Identities" with Young Children: Making Politics Visible 31
Glenda Mac Naughton, Karina Davis, and Kylie Smith

3 The Dynamics of Whiteness: Children Locating Within/Without 49
Karina Davis, Glenda Mac Naughton, and Kylie Smith

4 Intersecting Identities: Fantasy, Popular Culture, and Feminized "Race"-Gender 67
Glenda Mac Naughton, Karina Davis, and Kylie Smith

5 Masculinities, Mateship, and Young Boys: Defending Borders, Playing Footy, and Kissing Meg 85
Karina Davis and Glenda Mac Naughton

Part I References 99

Part II Exploring the Politics of Adults' Racialized Identities

6 (Un)masking Cultural Identities: Challenges for White Early Childhood Educators 113
Karina Davis

7 On "Race" and Resistance: Transforming Racialized Identities—A Personal Journey 127
Merlyne Cruz

8 Adults Constructing the Young Child, "Race," and Racism 139
Sue Atkinson

9 Languages Matter: My Subjective Postcolonial Struggle 155
Prasanna Srinivasan

10 Working Within, Beyond, and Through the Divides: Hopes and Possibilities for De-"Racing" Early Childhood 167
Sue Atkinson, Merlyne Cruz, Prasanna Srinivasan, Karina Davis, and Glenda Mac Naughton

Part II References 181

Appendix 187

Subject Index 193

Series Editors' Preface

This book makes a major contribution to our series, *Critical Cultural Studies of Childhood*, and to the literature in childhood studies and early childhood education. Editors and contributors confront an often-silenced dialogue in our field—that of the prominence of "race" in young children's sense of self and others, or the "race"-ing of early childhood, as the editors name such issues. This volume critiques pervasive assumptions of racial innocence and challenges readers to confront and better understand racialized identity politics in the lives of both children and adults in their lives. Chapters include work that centers Aboriginality and draws from postcolonial theory and decolonizing methodologies. Further, it offers points of possibility and hope for this work through accounts of educators who are reframing how they engage with "race" in their own lives and those children with whom they share classroom and life spaces.

Coeditors Glenda Mac Naughton and Karina Davis interrogate their white privilege and problematize current racial dynamics in Australia and beyond. Other contributors to the book include several of the team of scholars who have built an internationally visible scholarly record at the Centre for Equity and Innovation in Early Childhood at the University of Melbourne, which has long championed antibias curricula and in-depth studies of complex social phenomena. One of the striking aspects of scholarship associated with the Centre is its deep respect for children, focus on the complexities of diversity, and grounding in principles of social justice and human rights.

The lines of inquiry reflected in this volume—ways in which "race" is perceived and enacted by young children, alongside the persistent denial that children see "race," class, gender, or other of their and others' multiple identities—are significant ones. By using empirical research projects that have entered the life worlds of children in various creative ways, the chapters in this book report findings and theoretical framings that are extremely relevant to this book series and an international readership of scholars and graduate students interested in children's identities, role performance, and perceptions of others.

"Race" and Early Childhood Education: An International Approach to Identity, Politics, and Pedagogy addresses the visible gap in empirical research, both international and focusing on Australia, that explores the relevance of "race" to children and educators and examines the challenges, tensions, and responsibilities for white teachers and researchers looking at these issues. Readers will appreciate the ways in which contributing authors examine multiple and intersecting identities and draw from contemporary research and theories to problematize prevailing theories and assumptions. Drawing from an array of feminist, postcolonial, critical race and poststructuralist theories of identity politics, contributors foreground children's voices and raise provocative new questions for our field. The explicit call for new pedagogical possibilities for antidiscriminatory identity work in early childhood settings is compelling and we anticipate that this book will have an impact on our work with young people in many contexts and life worlds.

—Series Editors
Beth Blue Swadener, Marianne N. Bloch, and
Gaile Sloan Cannella

Acknowledgments

A book of this sort is reflective of work committed to and carried out by more than the editors and the authors named in it. First, the editors and authors would like to acknowledge all the participants of the research projects mentioned in this book. It is their willingness to share their words, thoughts, experiences, and understandings that make this work possible and their participation is appreciated. Second, the editors would like to acknowledge their colleagues at the Centre for Equity and Innovation in Early Childhood (CEIEC) in the Melbourne Graduate School of Education at the University of Melbourne. These colleagues have been, and continue to be, supportive and encouraging of the people and the work mentioned in this book and continually work to create and re-create an environment of collegiality and community so important in the sort of work we do.

Part I of the book draws predominantly from data from the PCCRCD research project. There were three active researchers in this project and Karina and Glenda wish to thank and acknowledge Kylie Smith in the work (both practical and intellectual) she contributed in this project. Her generosity and support as a colleague and her desire to work to problematize the truths in early childhood have made (and continue to make) working with her a delight.

Part II of the book draws from projects of a variety of members of the CEIEC and we wish to thank them for their desire and will to come together to construct this section and their generosity in sharing their words and experiences. Specifically, Sue brings a great intellectual knowledge base along with an irreverence for the institution that works to shatter the insulated and formal world of the academy and ground our belief in the importance of doing this work. She brings a wise and questioning presence to all of our work; Merlyne brings powerful stories of life and deconstruction while dancing her way through the CEIEC offices. Her quiet thoughtfulness, generosity, and support works to center much of the work that we have done here in this book and within the CEIEC; Prasanna brings a passion and belief in the work we all do with a willingness to laugh, cry, and shout at the structures that make this work difficult. She is always willing to discuss and debate and come together to explore and importantly, question, our ideas.

Any book comes together in the final stages only with the help of several people offering and providing all sorts of support. Thanks to Kate Alexander for making space in her busy days to teach Karina the intricacies and skills involved in being computer literate and to Patrick Hughes for generously editing and commenting on chapters at the very last minute. This helped enormously in getting the book finished on time.

On a personal note, Glenda wishes to thank Patrick for his ongoing emotional, political, and practical support and for sharing with such commitment over so many years the antiracist project at heart of her work in this book. Karina wishes to thank John for his ongoing commitment to finding ways in which we both get to do the things we want and need in the midst of negotiating our busy lives and to her Mum and Dad, Maree and Peter, for always being so generously willing to help create space and time when the struggle to balance it all feels too great.

Contributors

Sue Atkinson is an Indigenous Australian, a Yorta Yorta woman from Victoria who has worked in the early childhood field for thirty years. She was recently awarded her PhD for her thesis "Indigenous Self Determination and Early Childhood Education and Care in Victoria," and she is currently an Honorary Research Fellow at the University of Melbourne's Centre for Equity and Innovation in Early Childhood in the Graduate School of Education.

Merlyne Cruz is currently engaged as a Research Fellow at the Centre for Equity and Innovation in Early Childhood, Graduate School of Education at the University of Melbourne. She has worked in the field of education for many years covering a range of roles as a primary school teacher, an early childhood practitioner, as well as teaching in the University sector. Her research interests are in Anti-Colonial Thought, Indigenous Knowledges, Critical Spirituality, and Diasporic Identities.

Karina Davis is a Research Fellow and Lecturer at the University of Melbourne's Centre for Equity and Innovation in Early Childhood in the Melbourne Graduate School of Education. Her research examines how postcolonial and white identity theories can be used with early years educators to explore how they can examine, critique, and rethink their work with regards to cultural diversity. In her work in the CEIEC, Karina leads the research theme "Identities, Equity and Pedagogies for Social Change," and has been actively involved in leading and working with both young children and adults in projects focused on a range of topics including identity, curriculum, and policy.

Glenda Mac Naughton has worked in the early childhood field for over thirty years. She is currently a Professor in the Graduate School of Education, at the University of Melbourne where she established and now directs the Centre for Equity and Innovation in Early Childhood (CEIEC). Glenda has researched, published, and been involved in teaching and learning about social justice and equity issues in early childhood for the past twenty years. Two recently published books with CEIEC colleagues focus on action research in early childhood and young children as citizens.

Kylie Smith is a Senior Lecturer and Research Fellow at the University of Melbourne's Centre for Equity and Innovation in Early Childhood. Her research examines how theory and practice can challenge the operation of equity in the early childhood classroom and she has worked with children, parents, and teachers to build safe and respectful communities. In her work with the CEIEC, Kylie has been actively involved in leading consultations with young children in curriculum and policy making in the early years.

Prasanna Srinivasan has worked in the field of early childhood, as an early childhood practitioner for more than ten years. Prior to this, she had used early childhood services as a parent. Her journey began by questioning the norms and practices as a parent and then as a practitioner. She tried to engage with children, families, and staff with notions of social justice and equity and exploring varied paradigms to be able to do so. Her quest and struggle to enable all voices to be heard equally made her dissatisfied with practices that were attached to inclusion She is now a Researcher/Lecturer with the Centre for Equity and Innovation in Early Childhood (CEIEC) in the Graduate School of Education at the University of Melbourne, trying to rethink current ideas about inclusion, especially the inclusion of cultures using postcolonial theories.

INTRODUCTION

Thinking Differently: The Call and the Desire

Karina Davis and Glenda Mac Naughton

Not many people know much about us.
That's why I want to share some things with you.
Things about us. About our land.
Things you may not have heard before…
…What I remember most about
those times is that I was totally free.
The choices were always mine, as it
was with all the members of my family.

Randall and Hogan, 2008, pp. 9, 19

The Terrorists
They are everywhere
I wear paranoia
like armour
like stone
like a raincoat
when it rains
when it doesn't
when smothered
by their attacks
I want to die
I want to kill
the fucking bastards
for making me feel that

being born in Australia
and being an Australian
are not the same.

Chau, 2008, p. 26

The idea that young children are innocent or ignorant about "race" and, therefore, are incapable of acting with "racial" intent is challenged in this book. We present empirical research evidence from several early childhood communities in Australia that children's and educators' lives are clearly "raced"; and we draw on this research and on decades of other international research and writing (Brown, 1998, 2001; Cannella and Viruru, 2004; Dau, 2001; Derman-Sparks and ABC Taskforce, 1989; Freire, 1970, 2002; hooks, 1994, 2003; Lane, 2008; Mac Naughton, 2005; McLaren, 1997) to explore the "racing" of young children. We use the term "racing" of young children to capture the complex and active individual and institutional sociocultural and political processes that form young children's feelings, desires, understandings, and enactments of "race" in their daily lives.

"Race" is a highly emotive and contested term. To understand why we use the term "race" in this book and why we put it in inverted commas it's helpful to turn to its history. Historically, "race" as a term applied to people was created to signify and name hierarchical classifications of human bodies ("racial" groups) and to link each group in the classification with specific intellectual, moral, and cultural traits. These "race" hierarchies have always benefited the group at the top of the hierarchy and have led to the suppression and oppression of "other" groups in the hierarchy. Briefly exploring this history shows us how "race" did not always exist but came to exist at a particular point in time for particular reasons and with particular effects for different groups of people.

"Race" Born in the 1500s

"Race" was first used in the English language in the early 1500s (Ashcroft, Griffiths, and Tiffin, 1998). Its use developed in Europe in the 1600s and it came to clearly represent the process of classifying people into groups (Smedley, 1999). A paper by a French doctor and anthropologist François Bernier is reportedly the first paper to outline a scientific system of racial classifications for people. His paper was titled: "New division of Earth by the different species or races which inhabit it" (Bernier, 1864, cited in Gossett, 1963).

The first "race" hierarchies, such as Bernier's, arose at a time of European expansionism and colonization of large parts of the world. It drew on the idea of a "Great Chain of Being" (Bynum, 1975; Lovejoy, 1936) that placed people and things in a hierarchy of value with God at the top, angels next, followed by people (men, then women, then children), then animals (who had their own hierarchy) with rocks at the bottom of the chain. The higher in the chain a person, animal, or thing, the closer to God they were, and the more noble and more spiritually moral they were.

The early "race" hierarchies drew the ideas linked to the Great Chain of Being idea to place European colonizers closer to "God" than "other" "racial" groups. This was supported by publication of Charles Darwin's *Descent of Man* (1874) in which he argued that it was an evolutionary fact and necessity that stronger "races" replace weaker "races" and that European "races" were superior to other "races." These ideas combined to support and build the ideology of white "racial" superiority that made colonization possible. It was used to justify the oppression and repression of those they colonized. For instance, when the British colonized Australia in the 1700s, the Indigenous peoples of Australia were seen as significantly lower in the Great Chain of Being than the British colonizers. Aboriginal Australians were seen as primitives and savages whom the superior British "race" could and should civilize. This justified their invasion and possession of other people's land and their management (at first in the form of genocide, then protectionism and assimilation). In the name of civilizing came a long and often violent history of suppressing and oppressing Aboriginal people and their languages and cultures. The oppression of First Nations peoples in Canada and North American and the African slave trade were similarly justified with similar tactics of oppression and suppression enacted. From its birth, the United States was based on making clear "racial" distinctions between African Americans, European Americans, and Indigenous (Native) Americans that placed European Americans at the top of the hierarchy.

"Race" as a Biological Fiction

Over time and in different countries there have been various efforts to classify people racially from Coon's 1800s hierarchy of five "races" (Causasoid, Congoid, Capoid, Mongoloid, and Australoid) to Huxley's nine "races" classification system (Banton, 1998; Corlett, 2003). Several nation states from South Africa to Singapore to the

United States still have their own "racial" classification systems. The existence of these different classifications shows clearly that there is no single, universal way to categorize people according to "race." As Omi and Winant (1994) argue, "racial" systems of categorizing people are social and historical processes, rather than biological facts:

> Although the concept of "race" invokes biologically based human characteristics (so called "phenotypes"), selection of these particular human features for purposes of racial signification is always and necessarily a social and historical process. In contrast to the other major distinction of this type, that of gender, there is no biological basis for distinguishing among human groups along the lines of "race." Indeed, the categories employed to differentiate among human groups along racial lines reveal themselves, upon serious examination, to be at best imprecise, and at worst completely arbitrary. (p. 54)

Whilst the biological basis of "racial" classification systems is still debated, there is strong scientific evidence against this (see Blackburn, 2000; King, 1971; Zack, 1998). For this reason, "race" remains a complex and highly contested concept with a highly political history and present. Part of that history connects to how "race"-color terms such as black and white are understood and used.

In this book, we do not use black/white as simple descriptors of skin tone. Instead, we use these terms to acknowledge the politics that generate these colors as "race"-terms and that connect identities to historical and social constructions of whiteness and blackness in "race" classificatory systems over time. For those people who have been named and now choose to self-name as "black," the term is a term around which to politically organize and challenge racism. Many people with a variety of skin tones "own" being "black" because it offers a sense of belonging and political affiliation to civil rights struggles against racism in many countries across many times.

Turning Race into "Race" as a Social and Political Construct

Throughout this book we follow a long-standing political convention of placing the word "race" in inverted commas (e.g., Darder and Torres, 2004) to highlight its problematic origins and implications. We do not use "race" to categorize people and groups based on genetic traits but to acknowledge that "racial" and "racialized" identities have resulted from a history of actions and institutions in which "race" has been used to classify, name, oppress, repress, and silence specific groups of people. It also recognizes the struggles and/or privileges groups have faced in these processes as real, not fictional and as such as a concept "race" continues to have a real effect in people's lives. In using the term "race" it is still necessary to acknowledge this but also because it still constructs our relationships (personal and institutional) with each other. As Darder and Torres (2004) explain:

> This is apparent in racialized discourses of hierarchy, in which members of dominant groups assert their superiority over others, and in racialized discourses of solidarity, in which subordinated groups assert their unity and rights. As such, "race" may not be a biological fact, but it certainly is a social reality. (Darder and Torres, 2004, p. 5)

Subsequently, whilst "race" has been discredited as a biological and genetic fact (see Shipman, 1994) the effects of regarding "race" as a biological and genetic fact remain.

Changing "Races," Changing Contexts

In this book, as part of the antiracist work we do, we use the term "race" to remind us that racism still exists and needs to be challenged. Racist national histories continue to haunt the identities of many people in previous colonies; and many people continue to believe that "race" is, indeed, a biological and genetic fact, despite evidence to the contrary. While "race" and racism always have local features, they always appear in global contexts and they always have global histories. In "race" histories, the relationships between people in different parts of the world (e.g., Europe, Africa, and Australia) come to the fore. Colonization of one part of the world by another has been the colonization of one group of people by another. Geopolitics is the study of these histories and relationships (Mac Naughton, 2005).

Consequently, amongst contemporary geopolitics, antiracist educators must walk a fine line: they must confront local social relationships with their specific contexts and histories; but they must also acknowledge that those local relationships are inseparable from international histories and contexts. In Australia—as elsewhere—children and adults negotiate their social identities within particular local histories, politics, and understandings of "race" that both reflect and contribute to wider international histories, politics, and understandings. Consequently, while early childhood educators work within their specific local "racial" circumstances, they can collaborate with like-minded colleagues overseas who face equivalent (but not identical) local "racial" circumstances by sharing antiracist strategies that work, affirming the need for antiracist work and learning from the challenges and possibilities offered in each context.

Why Antiracist Pedagogies in Early Childhood?

Antiracist pedagogies involve teaching and learning to challenge racism. There is a long history of this work in many countries internationally (for example, see Brown, 2008 and Lane, 2008–United Kingdom; Dau, 2001–Australia; Derman-Sparks and ABC Taskforce, 1989 and Ramsey, 2005–United States; Murray, 2001–Ireland; Prott and Preissing, 2006–Germany; and Van Keulen, 2004–The Netherlands).

It is through antiracist education that we can begin to challenge prejudice, racism, and discrimination and unpack the sense of disconnection and superiority from others experienced by the dominant. Through questioning and challenging constructions of identity that are "raced" within this book, we strongly support the argument for antiracist education in early childhood to construct more hopeful and "racially" just and fair presents and futures for all children.

Our hope is that this book will help antiracist educators internationally to strengthen and renew their pedagogies by hearing how others have explored and understood children's and adults' local experiences and understandings of "race." We invite readers to explore similarities and differences between the politics of "race" in the lives of young Australian children and the children in their own contexts.

Looking closely as researchers at how "race" touches young children's lives in Australian early childhood settings has led us to understand identities as highly politicized. In this, we see both children and adults as intricately and integrally located within these politics. We use the research we share from our own work and that of other researchers internationally to argue that the early childhood environment contributes to the "race"-ing of young children. We use this phrase to capture the complex and active individual and institutional sociocultural and political processes that form young children's feelings, desires, understandings, and enactments of "race" in their daily lives.

It is our firm conviction, following many others (e.g., Derman-Sparks and Ramsey, 2006) that developing antiracist pedagogies requires educators to locate and name the effects and implications of "race" in children's lives and in their own lives. We hope that this book provokes and inspires this process for educators of color who have experienced racism as children and as adults. For those educators who are connected within powerful and dominant "race" identities we hope the book supports them to tackle the often difficult process of seeing their connection to "race"-power in supportive and productive ways.

The Book: A Snapshot

Part I examines the politics of children's "racialized" identities. It examines different ways to think about children, "race," and identity and their implications for antiracist pedagogies; it discusses different ways to understand the place of "race" in children's and educators' lives; and it explores how children construct and negotiate "race."

Chapter 1 reviews some key ideas and themes in research on children and "race" for the past eighty years and argues for a need to insert thinking about the politics of identity more firmly into this work.

Chapter 2 develops an argument about how the politics of identity can be understood and used to explore the dynamics of children's "racing" in early childhood settings. It also explores how methods from the "race" and children research can be used to do this in research and in early childhood spaces.

Chapter 3 explores the dynamics of whiteness in young children's words and texts and how educators can use texts beyond early childhood to do this too.

Chapter 4 examines the ways in which global popular culture icons such as Barbie are implicated in the "racing" of young Australian girls.

Chapter 5 looks at how masculinities and "race" intersect in the play of boys and make whiteness a position of privilege that is hard to deconstruct.

Part II focuses on the politics of adult's racialized identities. It explores how early childhood educators who confront "racing" from different positions and experiences can acknowledge and/or name "race" and how they can undertake antiracist pedagogies.

Chapter 6 highlights how white early childhood educators can work to acknowledge and locate their own white racial identities in order to engage with their connections and complicity with issues of "race" in their own lives and their lives with children and families.

Through a personal journey exploring and deconstructing "race," Merlyne Cruz in Chapter 7 examines the ways and possibilities that spaces can be decolonized in order to work in antiracist and anticolonial ways.

Chapter 8, through drawing predominantly on the voices of Victoria's Indigenous early childhood community, highlights how "race" is contested in the lives of children and families and how it is possible to work toward self-determination of Indigenous communities within early childhood.

Chapter 9 focuses on how dominance works to normalize through language and how linguicism exists in early childhoods spaces with the effect of supporting dominance and marginalizing others.

Chapter 10 is a conversational weaving of all contributors' responses to the chapters in this book in efforts to explore ways of working within antiracist pedagogies.

A Note on the Language and Ideas in This Book

This book has been written by a diverse group of authors. We are now each working as researchers in a university but we have each also worked more or less recently as early childhood educators with children and families. As researchers we are challenged to theorize what we do using language, theories, and concepts that are not often used outside of university contexts. We are also expected to write for fellow researchers familiar with the language, concepts, and theories we use. As antiracist educators we are challenged to find ways to talk about our research that makes sense to diverse audiences, many of whom do not engage regularly, if at all, with the language, theories, and concepts that have become part of how we each have come to think about "race," racism, and antiracism. That language is the language of postcolonial, critical race theory, postmodern, and poststructuralist scholars. Many of these scholars are philosophers, some are educators, others lawyers, and several antiracist activists. In their language we find ways of thinking and speaking that bring us hope, challenge, inspiration, and hard political thinking that has helped to spur our work to be antiracist educators and researchers.

For those of you unfamiliar with the language, theories, and concepts we use, we have tried to make them "knowable" and "useable." However, we each remember well our own struggles to find them "knowledge" and "useable." For those of you familiar with the language, theories, and concepts we use, we hope our use of them strengthens your own work. Our aim in writing this book is to support and strengthen the work of antiracist educators and researchers alike. We hope it offers insights into research on "race" and children that supports your own arguments about the need for antiracist education to begin early, inspires you to further this research, helps you shape new and/or stronger arguments for engaging with young children's "racing" and that it brings ideas that refresh you or help you start this work. Your own relation to "race" is likely to influence how you read what we have written—whether our work is affirming or deeply challenging—we offer it in a spirit of hope that it can firmly put to rest the idea that "race" is not the business of early childhood education.

A Note on Australia's Local "Racial" Circumstances

In our work with antiracist educators in early childhood settings in Australia, we have tried to walk that fine line between local and global. In this work we cannot ignore the history of "race" that links us to the British who colonized Australia and the other ex-colonies she colonized. "Race" relations in Australia today cannot be separated from colonial discourses of "race" that emerged in the 1700s to justify its colonization that has ever-present effects in contemporary relations between Indigenous and non-Indigenous Australians. We track these relations in what follows to offer a context for the Australian research we present in this book.

A History of Colonization and its Effects

Australia's long history of Indigenous citizenship was disrupted by the arrival of white European "settlers" in 1788. In the subsequent 220 years, "race" has sparked war, resistance, violence, and dispossession. As Atkinson (2008) states:

> Australia was colonised by the British in 1788 and by mid 1800s Indigenous Victorians had suffered mass, often violent dispossession and dislocation marked by the loss of land, culture and life itself. More than 200 years later the survivors of colonisation continue to salvage, strengthen and re-establish Indigenous culture, community and self-determination. (Personal communication)

The governor of Australia had been ordered by the British Government (1787) to negotiate the ownership and use of the land with its Indigenous inhabitants. Despite this, Indigenous Australians were removed forcibly from their lands and, in many instances, were attacked and even massacred. In the late 1800s and early 1900s, state and territory governments established Aboriginal Protection Boards, which gave the government and/or chief protector power to remove Indigenous children from their families into state "protection" and to remove Indigenous people from their traditional lands and force them onto and between reserves. In these reserves, Indigenous people's rights to freedom of movement and to practice their cultures and speak their languages were restricted severely and often violently. This persisted well into the twentieth century: in 1967, a referendum gave Indigenous peoples full citizenship and gave the Australian Commonwealth government the power to legislate for and on behalf of Indigenous people (Critchett, 1990; Haebich, 2005; Langton, Tehan, Palmer, and Shain, 2004; Parbury, 1999).

As Australia's State, Territory, and Commonwealth governments denied Indigenous people their human rights and struggled to contain Indigenous resistance, they also had to negotiate and manage immigration. In the eyes of the British colonizers, Australia was a white country (Langton, 2003) and Australians were proud, antiauthoritarian white (male) larrikins and battlers who survived their harsh conditions with good humor (Martin, 2004). Throughout the twentieth century, successive Commonwealth governments enacted laws—referred to collectively as the "White Australia policy"—that favored immigrants from Anglo-Celtic and European backgrounds. These policies were relaxed in the 1950s and 1960s, but were only abolished in 1973, after which immigration laws reflected governments' more egalitarian approach to migration. Nonetheless, the legislation encouraged migrants to assimilate into an undefined "Australian way of life." In 1978, the Commonwealth government introduced a policy of multiculturalism, celebrating everyone's right to retain and practice their own culture and language. However, despite this newfound respect for diversity, many migrants and Indigenous peoples were still encouraged to assimilate into the (still undefined) "Australian way of life" (Elder, 2005; Moran, 2005; Povinelli, 2002; Pung, 2008).

At the level of the Australian Commonwealth, multiculturalism flourished for many years until 1996, when a conservative government led by John Howard came to office. In the ensuing eleven years in office, Australia immigration policies focused increasingly on people who were deemed "undesirable" immigrants (Neumann, 2004). In October 2007 the minister for immigration, Kevin Andrews, announced that the government had decided to restrict immigration by black Africans because they "failed to integrate" into Australian society (Hobday, 2007). Populist rhetoric supported and encouraged verbal and physical attacks on people from South-East Asia and on Indigenous people. Despite Australia's agreement to accept refugees at levels derived from international agreements (e.g., the 1951 UN Convention relating to the Status of Refugees and the 1967 Protocol relating to the Status of Refugees) refugees were denied entry to Australia on the grounds that they were "illegals" and "queue jumpers"; instead, they were detained on small islands that were under Australian control but deemed to be "outside" of its border. These domestic policies were accompanied by a foreign policy based on allegiance to the United States and forging something of a neocolonial relationship with countries in the South Pacific. After the 9/11 attacks on targets in the United States, Australia, like others in the Western alliance, introduced antiterrorism laws under which police have held individuals suspected of terrorism, although when their cases came to court, several of them have been freed (*ABC News*, 2007; Dasey, Allard, and Marriner, 2007).

In one respect, however, the Howard government was no different to its predecessors in Australia's post-1788 history. Like them, it failed to improve the social and material circumstances of Indigenous peoples and, like them, it failed to ensure that Indigenous people can claim their human rights and determine their future. Indeed, the Howard government eroded Indigenous peoples' rights still further. It rejected the growing number of calls to apologize formally and on behalf of the Australian people to Indigenous peoples for the forced removal of their children to white orphanages and foster homes as part of a policy of forced assimilation between 1869–1970. It abolished the Aboriginal and Torres Straights Islanders Commission (ATSIC), a national Indigenous representative body established by the Commonwealth Government in 1989. It acted to amend the Native Title Act (1993) to disallow rulings in favor of Indigenous land rights. Finally, in its last term in office, the Howard Government established a military-based "intervention" into remote Indigenous communities, under which government agencies assumed control of the communities' lands and property and of selected individuals' incomes and expenditures (Grubel, 2007; Rothwell, 2007).

Late in 2007, the Australian Labor Party, led by Kevin Rudd, won a federal general election. At the start of the first sitting of the new national Parliament after that election, Kevin Rudd—as prime minister—apologized to Australia's Indigenous peoples for the "laws and policies of successive parliaments and governments" and for the "removal of...children from their families" (Prime Minister of Australia, 2008) and he promised to improve the opportunities, health outcomes, and lives of Australia's Indigenous peoples. Since the 2007 election, refugee laws have been altered, mandatory detention of refugees has eased, and refugee detention centers have closed. At the time of writing, the intervention into remote Indigenous communities remains, but is under review (*ABC News*, 2008; *Lateline*, 2008).

References

ABC News. (2007). Anti-terrorism, migration laws "need scrutiny." *ABC News*, July 23.

———. (2008). Rudd stands by the NT Indigenous intervention. *ABC News*, June 21.

Ashcroft, B., G. Griffiths, and H. Tiffin. (1998). *Key concepts in post-colonial studies*. London and New York: Routledge.

Atkinson, S. (2008). Personal communication. October 28.

Banton, M. (1998). *Racial theories*, 2nd ed. Cambridge: Cambridge University Press.

Bernier, F. (1864). A new division of the earth. *Journal des sçavans*. Translated by T. Bendyphe in "Memoirs read before the Anthropological Society of London," 1, 1863–1864, 360–364.

Blackburn, D. (2000). Why race is not a biological concept. In B. Lang (Ed). *Race and racism in theory and practice* (pp. 3–26). Lanham: Rowman & Littlefield.

British Government. (1787). *Instructions to Governor Phillip*. Sourced at www.foundingdocs.gov.au/resources/transcripts/nsw2_doc_1787.rtf on November 4, 2008.

Brown, B. (1998). *Unlearning discrimination in the early years*. Staffordshire UK: Trentham Books Limited.

———. (2001). *Combating discrimination. Persona dolls in action*. Staffordshire UK and Virginia, USA: Trentham Books Limited.

———. (2008). *Equality in action: A way forward with persona dolls*. Staffordshire: Trentham Books, www.person-doll-training.org

Bynum, W.F. (1975). The great chain of being after forty years: An appraisal. *History of Science*, 13, 1–28.

Cannella, G.S., and R. Viruru. (2004). *Childhood and (postcolonization). Power, education and contemporary practice*. London and New York: Routledge.

Chau, K. (2008). The terrorists. In A. Pung (Ed). *Growing up Asian in Australia* (p. 26). Melbourne: Black Inc.

Corlett, J.A. (2003). *Race, racism and reparations*. New York and London: Cornell University.

Critchett, J. (1990). *A distant field of murder. Western district frontiers 1834–1848.* Carlton, Victoria: Melbourne University Press.

Darder, A., and R.D. Torres. (2004). *After race: Racism after multiculturalism.* New York: New York University Press.

Darwin, C. (1871/1874). *The descent of man*, 2nd ed. London: John Murray.

Dasey, D., T. Allard, and C. Marriner. (2007). Terror suspect's case adjourned. *Sydney Morning Herald.* Online, July 14. Accessed March 31, 2009.

Dau, E. Ed. (2001). *The anti-bias approach in early childhood*, 2nd ed. Frenchs Forest, NSW: Pearson Education.

Derman-Sparks, L., and P. Ramsey. (2006). *What if all the kids are white? Anti-bias/multicultural education with young children & families.* New York: Teachers College Press.

Derman-Sparks, L., and the ABC Taskforce. (1989). *Anti-bias curriculum. Tools for empowering young children.* Washington, DC: NAEYC.

Elder, C. (2005). Immigration history. In M. Lyons and P. Russell (Eds). *Australia's history: themes and debates* (pp. 98–115). Sydney: University of New South Wales Press.

Freire, P. (1970). *Pedagogy of the oppressed.* New York: Penguin Education.

———. (2002). *Pedagogy of hope.* London and New York: Continuum.

Gossett, T.F. (1963). *Race: The history of an idea in America.* Oxford, England: Oxford University Press.

Grubel, J. (2007). Aborigines say Australia intervention racist: study. *Yahoo! News*, October 13.

Haebich, A. (2005). The battlefields of Aboriginal history. In M. Lyons and P. Russell (Eds). *Australia's history: themes and debates* (pp. 1–21). Sydney: University of New South Wales Press.

Hobday, L. (2007). Migrant groups defend African refugee integration. *ABC News*, October 3.

hooks, b. (1994). *Teaching to transgress: Education as the practice of freedom.* London and New York: Routledge.

———. (2003). *Teaching community: A pedagogy of hope.* London and New York: Routledge.

King, J. (1971). *The biology of race.* New York: Harcourt Brace

Lane, J. (2008). *Young children and racial justice: Taking action for racial equality in the early years.* London: National Children's Bureau

Langton, M. (2003). Introduction: Culture wars. In I. Anderson, M. Grossman, M. Langton, and A. Moreton-Robinson (Eds). *Blacklines. Contemporary critical writing by Indigenous Australians* (pp. 81–91). Carlton, Victoria: Melbourne University Press.

Langton, M., M. Tehan, L. Palmer, and K. Shain. Eds. (2004). *Honour among nations? Treaties and agreements with indigenous people.* Carlton, Victoria: Melbourne University Press.

Lateline. (2008). Senate inquiry hears NT intervention putting Indigenous children at greater risk. *Lateline*, April 30.

Lovejoy, A. (1936). *The great chain of being: A study of the history of an idea.* Cambridge, MA: Harvard University Press.

Mac Naughton, G. (2005). *Doing Foucault in early childhood studies: Applying postrstructural ideas.* London: Routledge.

Martin, S. (2004). Dead white male heroes: Ludwig Leichhardt and Ned Kelly in Australian fiction. In J. Ryan and C. Wallace-Crabbe (Eds). *Imagining Australia: Literature and culture in the new new world* (pp. 23–52). Cambridge and London: Harvard University Press.

McLaren, P. (1997). *Revolutionary multiculturalism. Pedagogies of dissent for the new millennium.* Colorado: Westview Press.

Moran, A. (2005). *Australia: Nation, belonging and globalization.* New York and London: Routledge.

Murray, C. (2001). *"EIST", respecting diversity in early childhood care, education and training.* Dublin: Pavee Point.

Neumann, K. (2004). *Refuge Australia. Australia's humanitarian record.* NSW: UNSW Press.

Omi, M., and H. Winant. (1994). *Racial formation in the United States: From the 1960s to the 1990s.* New York: Routledge.

Parbury, N. (1999). Terra nullius: invasion and colonisation. In R. Craven (Ed). *Teaching Aboriginal studies.* (pp. 101–128). St Leonards, NSW: Allen and Unwin.

Povinelli, E.A. (2002). *The cunning of recognition: Indigenous alterities and the making of Australian multiculturalism.* Durham and London: Duke University Press.

Prime Minister of Australia. (2008). Apology to Australia's Indigenous peoples. Sourced at http://www.pm.gov.au/media/speech/2008/speech_0073.cfm on November 3, 2008.

Pung, A. Ed. (2008). *Growing up Asian in Australia.* Melbourne: Black Inc.

Ramsey, P. (2005). *Teaching and learning in a diverse world*, 3rd ed. New York: Teachers College Press.

Randall, B., and M. Hogan. (2008). *Nyuntu ninti (what you should know).* Sydney: ABC Books.

Rothwell, N. (2007). Clare on the back foot. *The Australian*, October 27.

Shipman, Pat. (1994).*The evolution of racism: Human differences and the use and abuse of science.* Cambridge, MA: Harvard University Press.

Smedley, A. (1999). Race in North America: Origin and evolution of a worldview, 2nd ed. Boulder: Westview Press.

Van Keulan, A. (2004). Young children aren't biased, are they? In R. Prott and C. Preissing (Eds). *Bridging diversity—an early childhood curriculum.* Weimar & Berlin: Verlag Das Netz.

Zack, N. (1998). *Thinking about race.* California: Wadsworth.

PART I

Exploring the Politics of Children's Racialized Identities

CHAPTER 1

Discourses of "Race" in Early Childhood: From Cognition to Power

Glenda Mac Naughton and Karina Davis

Too Young to Know, too Young to Educate? Debating Children's "Racial" Innocence and Antiracist Education

> Race consciousness, is ... as far as observation goes, an acquired trait, quite as much as the taste for olives or the mania for collecting stamps. *Children do not have it.* They take the world of human beings in which they find themselves as part of the order of nature and respond to a black or yellow face as readily as they do to a white, depending upon the character and intimacy of the association. (Park, 1928, p. 16. Emphasis added)

> One of the most depressing aspects of prejudice is the early age at which it rears its head. (Hewstone, 1988, p. vii)

As those two quotes illustrate, the question of young children's "racial" identities and what young children know about "race" has a long history. Park's statement is from his 1928 paper, "The Basis of Race Prejudice." Park was an U.S. academic who was writing at a time when there was considerable theoretical and research activity about racial prejudice in the United States (e.g., Bogardus, 1925; Frederick, 1927; Reinhardt, 1928). However, this work primarily focused on studying racial prejudice in adults and college students (e.g., Young, 1927). Park's paper broke this pattern by focusing on the development of prejudice and racial identities in young children. Eighty years later, we are still researching and theorizing about "race" and young children—asking what children know, when they know it, and how they come to know it. The intervening eighty years have seen the production of hundreds of studies about the "racing" of young children; and most of it has fundamentally challenged Park's contention that young children are innocent of "race." (We use the term "racing" of young children to capture the complex and active individual and institutional sociocultural and political processes that form young children's feelings, desires, understandings, and enactments of "race" in their daily lives.)

The first challenge to Park's belief in children's "racial" innocence came in late 1929 by Bruno Lasker's book *Race Attitudes in Young Children* (Lasker, 1929). Lasker refuted the belief that young children are "color-blind" and without a "race consciousness," arguing instead that children can and do form racial attitudes from a young age. Subsequent research has reconfirmed Lasker's contention consistently, but the belief in children's "racial" innocence persists, breeding uncertainty about whether and when antiracist education should be part of young children's lives.

Lasker (1929) posed a firm link between U.S. children's knowledge of "race" and their social contexts. He argued that children's attitudes were formed by what adults taught them, by children's experiences of racial segregation in the United States, and by the profoundly biased nature of knowledge in the U.S. school curriculum of the time. Lasker drew specific attention to how education shaped children's knowledge of "race." He showed how history and biology and other "presumably scientific subjects" (1929, p. 373) merged fact and fiction to produce racist images and understandings of black people in the United States. For example, of history textbooks he wrote:

> The heroes of history may merge with those of legend and fiction, but the naked savage pictured, in contrast with a fully dressed white man, as representative of the Negro race will have produced a mental impression which returns as the word "Negro" is mentioned. (1929, p. 373)

Reviews of Lasker's (1929) book suggest that his thesis about children and "race" caught the interest of educators and academics alike. His work was reviewed in educational journals such as *The Elementary School Journal*, (Buswell, 1929) and the *Educational Research Bulletin* (Nelson, 1929) and in academic journals from diverse fields of study (e.g., *The Journal of Negro History* [N/A, 1929], *The American Journal of Sociology*, [Reuter, 1929], and *Pacific Affairs* [Reuter, 1929]). The reviews were generally positive, but some reviewers were critical of his methods. Lasker had studied adults' recollections of their childhood awareness of "race" and several of his contemporaries saw this reliance on subjective, retrospective recall of experiences as a major methodological weakness that undermined his conclusions.

Lasker's call (1929) to free education of its racial bias parallels contemporary calls to use education to challenge the growth of prejudice in young children. Further, debate continues about how best to research the "racing" of young children, how "racial" identities are formed, and what an educator's role is in this process. We will return to these debates at several points in this book. In this chapter, we focus on what researchers have learnt about the development of "racial" identities in young children and about the influence of "race" politics. Decades of research suggests that children form "racial" identities at a young age, and this poses major challenges and possibilities for educators seeking "racial" justice for all children.

Children's "Racial" Awareness—When Do Children Become Aware of "Race"?

> Despite the fact that there is a large body of research suggesting that ethnic and racial identity are related to a variety of outcomes, there is little consensus as to how best to conceptualize and operationalize these constructs....Among these concepts is the idea that individuals' attitudes regarding their ethnicity and race unfold systematically as part of a normative lifespan model of psychological development. (Yip, Seaton, and Sellers, 2006, p. 1504)

The first studies of the psychological development of young children's "racial" awareness and self-identification (Clark and Clark, 1939; Criswell, 1937, 1939; Horowitz, 1939) examined the extent to which preschool children noticed physical "race" markers such as skin tone, facial structure, and hair type. Children were presented with photos, drawings, and other artifacts (including dolls) chosen to represent a diversity of physical markers of "race"—primarily skin tone and hair color and texture—and children were asked if if they look like any of the people in those images. For instance, Clark and Clark (1939) presented children with line drawings of black boys, white boys, a clown, a dog, and a hen. Boys were asked which drawing looked like them; girls were asked which one looked like their cousin, brother, or playmate. Clarke and Clarke (1939) concluded that:

> The greatest (most significant) amount of development in self consciousness and racial identification occurs between the third and fourth years. (p. 587)

Clark's and Clark's 1939 findings that very young children are aware of "racial" difference as early as three years of age have been replicated across the decades using their techniques. Additional and increasingly sophisticated techniques have been used to explore children's level of "racial" awareness, but these studies still ask variations of Clark's and Clark's two broad questions. Presented with representations of people with obvious physical differences in skin tone, physiognomy, and/or hair type, children are still asked (in some form or another): "Who do *you* look most like?" and "Who do *they* look most like?" (Children may also be asked to explain their choices.) Children's responses to these questions are taken to represent their level of awareness of "racial" differences. If children match themselves or others accurately to a representation of a person of the same "race," they are believed to be "racially" aware.

Children's "Racial" Awareness: Why Do Young Children Notice "Race"?

Developmental studies of children's "racial" cognition assume a "normative lifespan model of psychological development" (Yip et al., 2006, p. 1504). Such studies map the changing nature of children's reasoning and perception about "race" in their earliest years, seeking to mark the onset of prejudice and its precursors (e.g., Richardson, 2005). U.S. researcher Mary Goodman is credited with the first description of the development in very young children of attitudes to "race" (see Lynch, Modgil, and Modgil, 1992). Goodman argued that such development occurred in three phases:

- Phase 1 (around 2–3 years of age): Racial awareness. Children first notice racial difference.
- Phase 2 (around 4–5 years of age): Racial orientation. Children first express attitudes (positive and negative) towards specific racial groups.
- Phase 3 (around 7–9 years of age): True racial attitude. Children first express complex, racially stereotyped attitudes and prejudice. (Goodman, 1964)

Subsequently, other researchers have extended Goodman's work. For instance, Katz (1973, 1987) developed a more detailed, eight-stage schema of the development of children's racial identity and attitude. Since the late 1960s, developmental studies of children and "race" have built stage-based explanations, known as

"Cognitive-Developmental Theory" (CDT) (Levy, West, Ramirez, and Pachankis, 2004, p. 43), of how children's "racial" identity develops; and their explanations have generally (and more or less explicitly) drawn on stage-based Piagetian (1954, 1968) theories of cognitive development (see van Ausdale, 2004). Cognitive Developmental Theorists believe that young children's developing cognitive capacities shape how they perceive the social and physical world and how they understand physical differences and similarities between people; and so their "racial" attitudes develop as their cognitive capacities develop. According to Levy et al. CDT suggests that:

> Children's attitudes toward racial and ethnic groups are influenced by their ability to think about group information in complex ways . . . children exhibit prejudice because they are not cognitively sophisticated enough to be open-minded and racially tolerant. (2004, pp. 42–43)

Illustrative of this approach is the early 1990s work of Bigler and Liben (1993) in which they studied the link between 95 young children's memory of stories that challenged racial stereotyping. In their conclusion they hypothesize on the link between a child's increasing cognitive capacities and a decrease in its racial stereotyping:

> In summary, the data from the present study are valuable in demonstrating the importance of both racial stereotyping and classification skills for information processing, and in showing the utility of investigating these mechanisms at the individual rather than at only the group level. These data are especially useful in this regard insofar as the more usual between-subjects variable of chronological age was successfully replaced with direct assessments of the variables that are hypothesized to account for the developmental changes in memory distortion, that is, decreased racial stereotyping and more advanced classification skills. (pp. 1516–1517)

Cognitive Developmental Theorists—following Piaget—believe that young children have an innate need to understand their world and that to do so, they sort and classify items in their surroundings. As they sort and classify, they think about items' attributes and relationships with each other. For example, they search for similarities and differences between items and sort them into categories. This builds their capacity to think logically and to make rules in order to understand how the world works. Children sort and classify people into groups in the same way that they sort and classify other items into groups: they search for similarities and differences between people and sort them into categories. Children may compare and group people according to many diverse characteristics, such as size, hair color, gender, clothes, whether or not they have a dog, and so on.

Cognitive Developmental Theorists have studied how young children use rules for identifying who belongs in which category (Hirschfield, 1995, p. 241). They have examined the rules that young children use to classify and sort people according to physical markers of "race" (skin tone, physiognomy, and hair type) and to other characteristics (positive and negative) of "race." Broadly, they have found that children as young as three years of age use physical markers of "race" to sort people into groups; and that children between four and five years of age begin to attach social and emotional values to their "racial" groupings. These children are more than just aware of "race"; they express preferences for certain "races" and against others. Young children's positive preferences for their own group (in-groups) and their negative views of other groups (out-groups) is seen to result from their cognitive incapacity to see similarities between "in-groups" and "out-groups" and individual differences within each group (see Aboud and Doyle, 1995; Black-Gutman and Hickson, 1996). For this reason, young children's early racial preferences are directly linked to their immature cognitive systems.

Cognitive Developmental Theorists, following Piaget, emphasize that sorting and classifying people (as with other objects) is an innate, natural, and, therefore, universal activity. CDTs further emphasize that children gain the necessary cognitive classification skills through thinking about similarities and differences within and between groups that they see at around eight years of age (Levy et al., 2004). However, while some children use markers of "race" to sort and classify people, others do not. For example, when children are shown photos of people with marked differences in skin tone and they are asked to comment on those images, some young children remain silent about differences associated with physical markers of "race" (e.g., skin tone and hair type). Their silence has been explained in different ways. For some cognitive researchers, their silence shows their "racial innocence" and the immaturity and instability of their concepts of "race." For others—paradoxically—children's silence about "race" shows that they are aware that "race" has social implications (see Killen, Margie, and Sinno, 2006; van Ausdale, 2004).

From a developmental perspective, older children should outperform younger children on recognizing "race," because greater maturity is meant to bring greater competence. However, developmental psychologists have found that young children categorize people according to "race" better than older children, while children's silence about "race" grows with age. Apfelbaum, Pauker, Ambady, Sommers, and Norton (2008) explored this "developmental anomaly" (Apfelbaum et al., 2008, p. 1516) by studying 101 middle-class children in primary schools in the United States. They found that the more children were aware of racial stereotypes, the less likely they were to classify people according to physical markers of "race." To explain their findings, Apfelbaum and his colleagues drew on Piagetian developmental theory. Piaget argued that as part of their cognitive development, children internalize social norms, from which Apfelbaum and his colleagues concluded that older children's silence may show their greater capacity to internalize social norms such as the unacceptability of racial prejudice:

> As a result, it may be that younger children who have not yet internalized such social conventions and concerns about appearing prejudiced are able to outperform older children on a task for which the acknowledgment of race is a relevant consideration. (Apfelbaum et al., 2008, p. 1516)

In other words, children's silence on "race" is not necessarily evidence that they lack knowledge of "race" or see it as irrelevant when sorting and classifying people. Instead, it may demonstrate their growing understanding that "race" and what is said or not said about it matters in their specific context; and so children's silence on "racial" issues may be linked to their maturing social cognition.

Yet this still isn't the whole story. The use of markers of "race" varies between children irrespective of their age. For example, U.S. researchers found that black children use markers of "race" differently from white children (see Lynch et al., 1992). If different children use "race" markers differently at different times, it is difficult to maintain the argument that children's "racial" identity and attitudes to "race" develops according to a developmental schema that is the same for all children at all times and in all places.

Who Am I? Which "Racial" Group Do I Belong To?

CDT researchers have studied whether and how children identify themselves with a specific "racial" group, and they have mapped the cognitive processes involved in developing a sense of belonging to a specific "racial" group. From a CDT perspective, children cannot develop stable "racial" identities until they are around eight years of age (e.g., Aboud, 1988; Bigler and Liben, 1993), and CDT researchers have mapped the ages and stages through which children develop mature, stable, and fixed "racial" identities. Their findings have been varied. In the 1960s, U.S. researchers found that young black children rejected the identity of "black," while young white children readily identified themselves as "white" (Morland, 1963). However, their preferences were associated with their circumstances (Simon, 1974).

More recently, UK researchers found that black and white children between five and seven years of age identified their ethnicity accurately and that they associated it with their family ancestry (Davis, Leman, and Barrett, 2007). In addition, black children identified ethnicity with skin color. The researchers found no differences in the children's self-esteem.

Moore (2002) suggests that for CDT researchers, any uncertainty, negotiation, or shifts in children's identity is evidence of their cognitive immaturity and incompetence:

> [They focus on]...the progressively sophisticated, stable and orderly individual acquisition of identity, such that the social negotiation of identity is viewed as immaturity and incompetence, especially as compared with adult identity. (p. 58)

From this perspective, children of color who identify themselves as white, or who see themselves as white in one context and not another, have immature or naïve "racial" identities and they lack an understanding of the rules of "racial" classification that comes with experience.

Who Do I Like? Who Do I Want to Be Like?: Children's "Racial" Preferences

Developmental psychologists have studied young children's "racial" preferences persistently (Brinson, 2001; Harrison and O'Neill, 2003; Johnson, 1992; Meltzer, 1939; Singleton and Asher, 1979; Spencer and Markstrom-Adams, 1990; Turner, Gervai, and Hinde, 1993; Zinser, Rich, and Bailey, 1981; Aboud, 1993). Their aim has been to uncover how children feel about different "racial" groups, including their own; and they have used techniques that are similar, sometimes identical, to those they have used to study other aspects of children's "racial" identity. These researchers have most often shown children images of people from different "racial" groups and then ask the children to indicate their preference for doing or not doing things with people from each group. Children's responses to these questions are believed to express their preferences for different "races."

Many contemporary studies use variations on questions asked in Horowitzs' (1938) early study of children's "racial" preferences. Horowitz and Horowitz (1938) showed children photographs of a range of people. To establish children's positive preferences, they asked them which people they would "like to sit next to at school," "play with," "want to sit next to at the show," "have for a cousin," "want to live near you," "want to come to your house for a long visit," and "like to take to town" (p. 313). To establish children's negative racial preferences, they asked them which people "live in a dirty house," "look stupid," "you do not like," and "your folks would not let you play with" (p. 313). A review of the children and racial preference research (Brand, Ruiz, and Padilla, 1974) identified several additional measures of children's "racial" preferences:

> Ranking scales...cross-ethnic comparisons on personality assessment devices, analysis of sociometric interactions, observation of intergroup behavior, attitude bias in disguised measures, and measurement of autonomic changes. (p. 860)

The results of several studies in the United States in the 1930s (including Clark and Clark, 1939 and Horowitz, 1938) have been repeated in subsequent studies. Broadly, researchers (mainly from the United States) have found that at around three years of age, black children are biased against their own group and prefer whiteness, while white children prefer their own group and are biased against other groups (e.g., Kircher and Furby, 1971; Stevenson and Stewart, 1958; van Ausdale and Feagin, 2001; Zinser et al., 1981). More specifically, black children consistently choose white dolls and ones that look pretty to play with (Gopaul McNicol, 1995), they prefer lighter skins (Averhart, 1997), and their "racial" preference is whites (Johnson, 1992; Kelly, 1995).

Most of this research has been conducted in the United States and has focused on the white majority group's attitudes to black Americans. However, researchers have found a positive bias to "whiteness" and a negative bias toward "blackness" in several ex-British colonies including Hong Kong, New Zealand, South Africa (Morland, 1969), Australia, (Palmer, 1990); in Germany (Best, Field, and Williams, 1976), France and Italy (Best, Naylor, and Williams, 1975); and in

Japan (Iwawaki, Sonoo, Williams, and Best, 1978). Some researchers have moved beyond the "black/white" binary that has dominated research in the United States. For instance, Australian researchers have found that ten- to twelve-year-old Anglo-Australian children have specific negative stereotypes of Vietnamese-Australians and of Italian Australians (Nesdale, 1987).

Are Very Young Children Really "Racially" Prejudiced?

In general, researchers within the broad tradition of CDT have linked children's expression of racial preferences strongly to the onset of prejudice; and they have tried to discover when children first notice "racial" differences and when they first express "racial prejudice" in order to prevent the onset of racism. There has been a consensus that children's views of "race" are immature and unstable until around eight years of age, and, therefore, they don't express "real" prejudice (van Ausdale, 1996). However, this consensus has been disturbed recently by a growing debate in the psychological literature about the extent to which young children's "racial" preferences express "racial" prejudice. In the 1990s, several researchers argued that young children can and do demonstrate "racial" prejudice (Hirschfeld, 1995), and that they may do so as early as three years of age (Black-Gutman and Hickson, 1996). However, to what extent do children's "racial" preferences express "racial" prejudice?

Social Identity Development (SID) theorists (e.g., Quintana and McKown, 2008) have addressed this question by drawing on the social psychology of intergroup relations (e.g., Davis et al., 2007; Kowalski, 2003; Cameron, Alvarez, Ruble, and Fuligini, 2001). They have examined how children build a sense that they belong to some social groups (e.g., a gender group, an ethnic group, a family) and how this affects their attitudes to other social groups. SID theorists have argued that children build a sense of belonging to a social group in three steps: classifying people into social groups; differentiating between groups regarded positively and those regarded negatively; and building self-esteem through membership of a social group (an "in-group") that is regarded positively.

Further, SID theorists have argued—like CDT researchers—that "racial" prejudice does not appear until seven or eight years of age because children's "racial" identity develops as part of their growing cognitive capacity to build a sense of belonging to a social group.

SID theorists have argued that children develop a "racial" identity in four phases, summarized by Davis et al. (2007) thus:

- Phase 1. Undifferentiated racial identity. Children under three years of age do not classify social groups using markers of "race," because they regard physical markers of "race" as insignificant to who they are and who others are.
- Phase 2. Racial awareness. Around three years of age, children notice physical markers of "race" and they begin to use those markers to classify social groups and to identify themselves.
- Phase 3. Preference for one's "in-group." Around four years of age, children begin to compare "racial" groups. They prefer to be with their own, because membership is a source of self-esteem; but they do not necessarily dislike people from other groups.
- Phase 4. Prejudice against "out-groups." Around seven years of age, children begin to dislike "out-groups," especially if they identify strongly with their "in-group" and if they regard an "out-group" as a threat.

As Davis et al. (2007) explained, the four phases assume that children desire to associate positively with their in-group; and that they regard their in-group as superior to "out-groups":

> Drawing on theorising from Social Identity Theory (Tajfel, 1978; Tajfel and Turner, 1979, 1986), social identity development theory (Nesdale, 1999a, 2004) proposes that children's racial and ethnic attitudes develop through a sequence of four phases, and that the development is primarily driven by social motivational factors such as the pursuit of positive distinctiveness of the in-group, self-esteem and a child's identification with the ethnic or racial in-group. (p. 515)

From an SIT perspective, the "racing" of young children is a social *and* a cognitive phenomenon. It depends not just on an individual child's developing cognitive capacities but also on their membership of some specific social groups (in-groups) and their relationships between these groups and others (out-groups). SIT's interest in how social relationships affect the "racing" of young children, while noteworthy, is not new. The earliest 1920s and 1930s research in the United States on the "racing" of young children was prompted by researchers' concern about the effects of racism in the wider society on children. For instance, Meltzer (1939) saw, "race attitudes as a social problem" (p. 104) warranting further research; and in the 1970s, researchers in the United States asked whether and how children's experiences of desegregation affected "race" relations between children in U.S. elementary classrooms (e.g., St. John and Lewis, 1975).

More contemporary work on young children and "race" has combined social learning perspectives with CDT perspectives to explore the impact of diverse environmental factors on what young children know, feel, and believe about "race" (see Aboud and Amato, 2001; Aboud and Doyle, 1996; Killen, Margie, and Sinno, 2006; Levy et al., 2004). This work represents part of a broader trend within the social sciences to more complex and nuanced understandings of the complex nature of prejudice, discrimination, and racism in young children's lives. The next section focuses on exploring the place of politics in furthering understandings of "race" in young children's lives.

Developing Cognition, Social Relationships, and Politics: Are Children Innocent of the "Racial" Politics of the Adult World?

(Mac Naughton, 2001a) has argued that social-cognitive and cognitive developmental studies of "race" and young children leave three important questions unanswered. The first question is: Why do so many children use skin color to sort and categorize people and to understand their relationships with their peers? In a study of a group of young Australian children, Mac Naughton (2001a) used dolls in research interviews and in free play sessions. She found that the children sorted the dolls by skin color and other physical attributes, rather than by equally obvious differences such as gender or clothing:

> Some children did not use skin color, but many did. What makes physical characteristics—such as skin color—that have historically been named by European colonial powers as "racial" so prominent in young children's classifications? Spivak (1990, p. 62) calls this process of basing decisions on skin color "chromatism." Why is "chromatism" a part of how many children understood who they were and who can be Australian? How does "chromatism" intersect with how young immigrant Australian children can and do construct their identities as Australians? (Mac Naughton, 2001b, p. 35).

Mac Naughton's (2001b) second question is: Why do Anglo-Australian children classify themselves with certainty as "white" (a politicized term that "racializes" skin tone), when other children name their skin color using words such as "pinkish" or "sort of light, gray-brown"?

If classifying skin tone is an impersonal cognitive process, as cognitive psychologists believe, then how do some children learn to use a term ("white") that is a highly emotive, politicized signifier of difference, rather than an impersonal descriptor of skin tone? Mac Naughton's (2001b) final question is: Why are young white children uncomfortable with darker skin?

Mac Naughton (2001a, 2001b) argued that social-cognitive and cognitive developmental studies of "race" and young children fail to answer these three questions because they fail to acknowledge the "racialized" political pasts and presents that make it possible for young children to classify and compare "races"; and that make some "racial" group memberships more desirable than others. These "racialized" pasts and presents include the voices and discourses of "race" that are the result of colonialism and white supremacy (Hage, 1998). Young children's ability and desire to use "race"-color to sort, classify, compare, and assign status to people is a direct result of the politicizing of skin tone in a specific country or nation state at a specific point and over time. This book seeks to insert, rather than edit out, the politics of the "racing" of young children. We argue that we should place children's "racial" understandings, preferences, and awareness within the specific politics of "race"-color classifications, comparisons, and status assignment in their particular circumstances; and that if we don't, we will silence the racism that produces these "racial" understandings, preferences, and awareness and we will miss the effects of whiteness in young children's lives.

In arguing the need to highlight the political and sociocultural contexts of young children's "racing," we echo the views of other contemporary researchers in the area, including Trew (2004), Northern Ireland; Targowska (2001), Australia; Connolly (2000), England. Our argument isn't new: Horowitz and Horowitz (1938) argued that it is as important to understand the social and political context in which children learn about "race" as it is to understand the cognitive processes through which that learning happen. We would add that just as children's understanding of "race" grows from their social and political contexts, so does their understanding of "race" hatred; and we agree with O'Loughlin (2001) that this is why we must challenge the idea that children are innocent or ignorant of "race":

> The one thing we do not want to do, I think, is assume that subject formation is inconsequential, or that we need to do nothing because

> the inherent innocence of children will protect them from performing hateful acts. (p. 63)

In the next chapter, we explore how to use postmodern and postcolonial perspectives on the politics of identity to acknowledge the politics of young children's "racing" in early childhood spaces and to work with young children to build identities that are "racially" just.

CHAPTER 2

Exploring "Race-Identities" with Young Children: Making Politics Visible

Glenda Mac Naughton, Karina Davis, and Kylie Smith

Putting Politics into Researching Young Children's Racial Identities

Politics is about power—the power to define the world and act in it and on it. Power is the capacity to exercise influence in the world about what is doable, permissible, desirable, and changeable in your lives. Inserting politics into the "racing" of children leads us to explore how power operates in young children's lives as they construct their "racial" classifications, comparisons, preferences, and status assignments. It requires us to ask how children and others exercise power in defining and influencing what is doable, permissible, desirable, and changeable in relation to "race" in their lives. Three ideas about identity are central to putting politics into researching young children's identities in order to acknowledge the complex dynamic of the social contexts in which they live, learn, and produce their racialized lives:

- Identity is chosen not fixed; it is therefore changeable.
- Identity is formed in and through discourse and therefore identity choices are limited or made possible through discourse.
- Identity is actively performed, not passively given.

We will chart what these ideas mean for understanding the "racing" of young children and then explore how research techniques such as observation, self-portraiture, child interviews, story, and ethnographic feedback can help locate the politics of "race" in young children's early education. We also look at how educators could pedagogically adapt our research techniques to locate the politics of "race" in early childhood spaces.

Rethinking Identity: From Fixed and Stable Individual Identities to Choosing Identities

As discussed in chapter 1, developmental approaches to identity see a stable "racial" identity as a sign of cognitive maturity. We particularly focused on CD and SID theories of racial identity development in chapter 1 due to their influence within the field.

However, beyond CD and SID debate surrounds the idea that a stable "racial" identity expresses maturity (e.g., Moore 2002). Several contemporary researchers argue that racial self-identification and categorization is fluid across the life-span, shifting with context rather than age (e.g., Hitlin, Brown, and Elder, 2006; Skattebol, 2005; Taylor, 2005). They emphasize the complexity and dynamism of "racial" identity, as it is socially mediated and constructed in specific contexts. For example, Skattebol, studying young children's identities in early childhood settings, sees identity "as a state of being or as a set of practices that position subjects in social space" (Skattebol, 2005, p. 192). This view is supported by UK research on the gender and "race" identities of young five- and six-year-old boys in a multiethnic primary school (Connolly, 2006). Over a two-year period, on an average of three days per week, the researcher took detailed observations of boys in various contexts during the school day. He observed the boys in the classroom, during playtimes, and during lunchtimes. He also interviewed the boys in small friendship groups. From this detailed and in-depth work he concluded that:

> Young children are actively involved in developing, negotiating and reproducing their gendered and racialised identities. It is clear from the examples that the 5- and 6-year-old boys discussed here have not simply been passively socialised into their gender identities and nor are they just uncritically repeating racist ideas they have picked up from elsewhere. Rather, they are playing a central role in appropriating and reworking existing ideas and also creating new ideas in response to the situations they find themselves in. (Connolly, 2006, p. 150)

Researchers such as Connolly (2006) who have explored children's identity in these in-depth ways over time in situ in early childhood settings argue that children have fluid, socially mediated "racial" identities. Young children strategically choose to self-identify "racially" in different ways in different contexts. Similar findings occur with adults and youth. For example, in recent research with adolescents it was found that the expression of a positive self-concept among American Indian youth was dependent on who they were with and where they are (Whitesell, Mitchel, Daufman, and Spicer, 2006). Alongside this, an increasing number of people are choosing to self-identify as American Indian in the US Census and the numbers cannot be explained by birth rates. People are choosing to "ethnically switch" identity (Huddy, 2001).

Identity switching offers a challenge to SIDT and CDT. Social identity theorists assume that individuals internalize socially predetermined and assigned group membership (Huddy, 2001) and then build their in-group preferences based on these. For CDT, identity switching in young children is read as a sign of naïve thinking based on their incapacity to accurately classify and compare social groups. In contrast with SID and CDT, postmodern/poststructuralist theorists see identity switching as a sign of human agency and the capacity of humans to play an active role in constructing themselves. For the "post-theorists," identity is not linked to maturity but to discourse. For those unfamiliar with it its use and understandings within postmodern/poststructuralist theories of identity understanding discourse and how it produces intimate relationships with power/knowledge, subjectivity and institutions is an important backdrop to the remaining chapters in this book. For this reason, we shall now look at how poststructuralists/postmodernists understand discourse.

Discourse, Identities, and the Limits to Identity Choices

For postmodern/poststructuralist theorists of identity, identity choice is produced in and through discourse. Discourse in this context refers to the frameworks we use to make sense of the world intellectually, politically, emotionally, physically, implicitly, and explicitly. They are manifest in how we structure institutions and social life. This understanding of discourse draws from the work of French philosopher and social theorist Michel Foucault (see Mac Naughton, 2005). For Foucault, discourse forms us. Discourse is more than words and ideas, it is ways of knowing and being in the world about what is doable, permissible, desirable and changeable at a specific point in history, in a specific context. Discourses operate through texts (images and words) and so are learnt in and through language. Children are born into a world of preexisting discourses and as they learn language they learn discourse. They become subject to discourses and discourses form their subjectivities (ways of giving meaning to themselves and their worlds). Identities, including racial identities, are therefore shaped in and through discourses of "race" that preexist the child's entry to the world. Skattebol (2005) illustrates this perspective well in her discussion of a group of children in an Australian preschool who strategically negotiate storylines that include and exclude on the basis of "race" by calling on discourses of whiteness to see baddies and "black" and "robbers" (p. 192).

However, children do not merely absorb discourse or copy it, they actively make it their own. Connolly (2006) reports convincingly on young boys' capacity to generate their own "racial" stereotypes that are specific to their local context. A stereotype about Asian boys generated by the five- and six-year old boys in his study in a UK multiethnic primary school was that "Asian boys cannot 'run fast'" (p. 150). He explains how this stereotype is not readily explained without acknowledgement of young children's capacity to actively construct and create meaning in their world:

> Such stereotype was not one that tended to be used within the context of the local estate and thus was not one that the boys had simply heard elsewhere and were now just repeating uncritically. Rather, the stereotype was generated by these boys specifically in the context of the school and the games of football they engaged in and controlled. In this sense, the boys' thinking within this context was partly mediated by football, which then influenced and shaped the way they were able to make sense of and justify their exclusion of Asian boys. (Connolly, 2006, p. 150)

Whilst children, such as those in Connolly's study (2006), work hard to make discourses their own and shape their own meanings in the world, they cannot shape them outside of discourse (see Mac Naughton, 2005). Poststructuralist and postmodern perspectives on identity see identity mediated, chosen, and performed within discourse rather than given and fixed (Hage, 1998 ; Mansfield, 2000; Pease, 2004). Children access these discourses in their daily interaction with other people, texts, and institutions. Early childhood spaces are intimately involved in offering children discourses through which they might "racially" construct themselves and enact their "race" relations with others. It is not just from the wider community that children access discourses of "race." Connolly (2006) clearly shows how the school mediates the gender and "racial" identity possibilities that white Anglo boys consider using in a multiethnic school in United Kingdom:

> It therefore makes little sense to see the boys' violent and racist behaviour as rooted in and being simply a product of the local community. The dominant forms of masculinity found within the school are as much mediated by the school itself as they are by the local community. In the present case study it was shown how the school tended to exacerbate and reinforce these forms of masculinity among the boys. (p. 150)

Foucault suggests that discursive fields form when multiple, contradictory discourses are in competition to shape what we do, desire, permit, and change in our social relationships, processes, and institutions (see Mac Naughton, 2005). They shape our subjectivities and, therefore, they shape what we believe is doable, permissible, desirable, and changeable for ourselves and for others. From this perspective, how we think about "race" is shaped in and through multiple and often contradictory discourses about what we should do, desire, permit, and change in our social relationships, processes, and institutions to produce a more "racially" just world. For instance, discourses of whiteness can shape what both white and black children believe they should do, permit, desire, and change in their relationships with each other. Should they talk to each other, permit friendships between them, desire a friendship with each other, or change friendships with others to make their friendship more possible?

Research in early childhood contexts using this perspective tend to be broadly ethnographic in orientation, studying how children use "race" (see Moore, 2002) to produce and use tactics of inclusion and exclusion (e.g., Connolly 2006; Mac Naughton, 2003; Skattebol, 2005; Taylor, 2005). These researchers ask questions of the "racing" of young children that shift attention from measuring "racial" attitudes to questions attending to children's lived experiences of "race." Such questions include:

- How do children enact and perform their racial identities?
- How do children enact "racial" privilege in their daily relationships?
- How does whiteness operate in children's lives to produce dynamics of inclusion and exclusion?

These questions assume that children cannot exist outside of a "raced" discursive field. However, they also assume that children's engagement with their localized discursive field will shape children's "racing."

In chapters 3, 4, and 5 we use research with young Australian children to show young children can and do make strategic choices in relation to their "racing." We map how the discursive field of "race" in Australian early childhood spaces is shaped by global discourses of "race," historical discourses of "race," and contemporary discourses of "race" that intersect with discourses of gender and class to produce what children believe is "racially" doable, permissible, desirable, and changeable. We show how children choose particular discourses of "race" and use silence as a political strategy to make choices about how they "racially" self-identify and identify "others." Silin (1999) explains how silence can be used strategically to different ends:

> Silence can signal resistance as well as oppression, voice can create new moments for social control as well as for personal efficacy. And words are notorious for concealing and transforming as well as revealing the truth of our lives. (p. 44)

As outlined in the Introduction, we also draw on postcolonial and critical race theories (Delgado and Stefancic, 1997) specifically critical theories of white identities (Frankenberg, 1993), to engage with the discursive dynamics of children's "racing," especially their "racial" silences and the meanings of those silences.

Inserting the Postcolonial and Theories of White Identities into the Politics of Identity

Poststructuralists argue (e.g., Butler 1997; Davies 2001) that the processes of being and becoming are full of tensions, as children gain access to competing discourses of gender, "race," and class and negotiate possibilities for themselves as gendered, "racialized," and classed beings. Amongst this, children attempt to clarify which forms of becoming are possible and desirable. In this process, children identify with particular ways of thinking and being, and resist, reject, or "disidentify" (O'Loughlin, 2001) with others. O'Loughlin (2001) examines how these processes connect to "Othering" in relation to the development of racial identity formation for children of Caucasian heritage and concludes that Othering for non-Caucasian children may be more important than self-identifying positively as Caucasian:

> While it would seem, intuitively, that identification with one's own racial or ethnic group is essential to identify formation, some writers suggest that, at least for Caucasians, the development of a white racial identity may depend as much on defining an Other that they are not, as on defining some essential characteristics of whiteness with which to identify. (p. 50)

In this context, Othering is understood as a process of seeing oneself positively by seeing an "other" as undesirable and lesser. It derives from hierarchical "us" and "them" thinking in which "them" is seen through negative stereotypes that may be based on "race," geography, sexuality, gender, ethnic, economic, religious, or ideological differences. "They" are therefore "lesser" to us. Historically, the process of Othering has been used to disempower, oppress, and colonize Indigenous people and other groups who were constructed through discourses of "race" as "racially" other and lesser to white European colonizers (see Mac Naughton, 2005).

Postcolonial scholars combine their interest in the historical and ongoing effects of colonization with postmodern/poststructuralist theories of identity to trace the operation of racist discourses of Othering (Ghandi, 1998) in adult and child identities. They share with postmodern theorists the idea that knowledge (e.g., of "race") is discursive and thus intimately connected with how power operates between different groups (Ashcroft, 1994; Bennett, 1998; Young, 1998).

Critical race theorists extend this thinking further by specifically focusing on how discourses of whiteness intimately link with discriminatory social structures to create the illusion that white individuals and communities can choose whether and how to engage with issues of "race," discrimination, and citizenship. These discourses allow white people to ignore or deny their part in individual, structural, and institutional discrimination. As stated above, in chapters 3, 4, and 5 we pay particular attention to how discourses of whiteness are used strategically by children to position themselves within specific discourses of gender, "race," and class as they search for the doable, permissible, desirable, and changeable with others and for themselves. In this endeavor we draw several research methods originating in the children and "race" research but we have adapted and combine them for use in an extended ethnographic study of young children's "racing" in early childhood settings (refer appendix for project details). In our adaptations, we have paid attention to previous critiques of these methods. We now briefly touch on key methods that have been used to research children and "race" and critiques they have generated as a context for the coming chapters.

Engaging Critically with Methods for Researching Children's "Racial" Identities

The Early Studies

In late 1929 when Bruno Lasker published his book *Race Attitudes in Young Children* (1929), he drew considerable methodological criticism of his work from fellow researchers in the field because he had studied adults' recollections of their childhood awareness of "race" rather than children themselves. Thus began a continuing debate about how best to study the "racing" of young children. Whilst the key protagonists in this debate have been psychologists and sociologists with contrasting research interests, questions, and methods, there has been a blurring of those disciplinary distinctions from the earliest research. For instance, Meltzer (1939) was an advocate of mixed-methods studies that used psychological methods to develop attitude measurement instruments and sociological methods to select diverse group subjects to show the complex interaction between children's religion, "race," class, and geographical location and their racial attitudes. Horowitz and Horowitz (1938), a contemporary of Meltzer, was clear that understanding the social context in which children learnt about "race" was as important as understanding the internal cognitive processes that made that learning possible. In many ways those earliest researchers exploring the "racing" of young children called for attention to its political complexities in much the same way as we are calling attention to it.

Establishing Trajectory and Traditions of Researching Children's "Racial" Awareness, Preferences, and Identification

The first psychological studies of preschool children and "race" explored children's racial awareness and self-identification (Clark and Clark, 1939; Criswell, 1937, 1939; Horowitz, 1939) by studying the extent to which children thought they looked liked photos, drawings, and other artifacts (including dolls) that had been chosen to represent a diversity of physical markers of "race"—primarily skin tone and hair color and texture. In studies across the decades, children have been shown images of people meant to represent different "racial" groups and asked to indicate their preference for doing or not doing things with those people. Variations on the questions in Horowitz's (1938) study persist in more contemporary studies. To establish positive preferences, Horowitz (1938) asked children to show the researchers those people that they would "like to sit next to at school," "play with," "want to sit next to at the show," "have for a cousin," "want to live near you," "want to come to your house for a long visit," and "like to take to town" (p. 313). To establish children's negative racial preferences he asked them to show the researchers the photos of people who "live in a dirty house," "look stupid," "you do not like," and "your folks would not let you play with" (p. 313).

Since the Clarks' original (1939) study, researchers have also used dolls to probe children's "racial" understandings, self-identification, and preferences. Generally, children are presented with dolls designed to physically represent specific "racial" groups and are asked a series of descriptive and evaluative questions. The Clarks' original tasks with the dolls were:

1. Give me the doll that you like to play with. (To determine "racial" preference.)
2. Give me the doll that is a nice doll. (To determine "racial" preference.)
3. Give me the doll that looks bad. (To determine "racial" preference.)
4. Give me the doll that is a nice color. (To determine "racial" preference.)
5. Give me the doll that looks like a white child. (To measure "racial" identification of others.)
6. Give me the doll that looks like a colored child. (To measure "racial" identification of others.)
7. Give me the doll that looks like a Negro child. (To measure "racial" identification of others.)
8. Give me the doll that looks like you. (To measure "racial" self-identification.)

Several researchers have used these same questions (e.g., Feinman, 1979; Stevenson and Stewart, 1958), others have adapted them. For instance, Simon (1974) asked the children in his study to point to the doll that:

1. You like to play with the best
2. Is a nice doll

3. Looks bad
4. Is a nice color
5. Looks like a white child
6. Looks like a colored child
7. Looks like a Negro child
8. Looks like you (Simon, 1974)

From Dolls to Interviews with Children about Friendship Preferences

Variations on the interview question, "Who do you like to play with best?" has been widely used to establish children's attitudes to other "racial" groups. For instance, Singleton and Asher (1974, 1979) asked children who they had their "best friendship" with; other researchers have explored with children who they would share with and who they preferred to do things with in different situation (Zinser, Rich, and Bailey, 1981).

Reviewing the children and racial preference research, Brand, Ruiz, and Padilla, (1974) identified several additional methods that have been used to measure children's racial preferences:

> Attitudes as measured by ranking scales..., cross-ethnic comparisons on personality assessment devices, analysis of sociometric interactions, observation of intergroup behavior, attitude bias in disguised measures, and measurement of autonomic changes. (p. 860)

One sociometric tool that researchers have used is the *Intergroup Contact Assessment*. It was used to measure contact between European Americans and African Americans. Participants were asked to look at a series of photographs of people and asked which group looked most like the people in their town, neighborhood, school, teams or clubs, friendships, and their family. Their responses were analyzed to establish their level of contact with the "other" (intergroup contact) and their level of bias in interracial peer encounters. A key finding was that Anglo-American children's level of bias toward other groups was associated with their level of contact with those groups—the less contact children had with another group, the greater their bias toward that group (McGlothlin and Killen, 2006).

Across the decades, changing critiques have been mounted against several of the more common methods that were used by researchers to examine young children's "racial" identities (van Ausdale and Feagin, 1996; Vaughan, 1986). First, there has been criticism that the dolls offered to children, certainly in the early doll studies, did not accurately portray physical differences between white and black

Americans. The researchers used white dolls painted black to represent black Americans. The criticism is that this offered children no real or meaningful way to choose between the dolls. Hence, black children's failure to self-identify with the black doll may have resulted from the lack of connection that they felt with a "black-up" white doll. Second, the doll tests and associated interview questions have often forced a binary choice on young children—the doll or person is either good or bad, someone you like or someone you don't. This forces choices that a child may not necessarily make in his/her daily life. For instance, they may not like either doll but are forced to choose one that they like. Third, in the force-choice method of interviewing children, researchers rarely probe children about why they make the choices that they do. This means that researchers may overinterpret or misunderstand the reasons why the children make the choices that they do. It may or may not arise from their "racial" understandings. Fourth, many of the studies on children and "race" have focused on measuring children's racial attitudes at a particular moment in time. These moments don't and can't capture the complex or dynamic nature of children's thinking and thus may provide an inaccurate sense of what they can and do feel and think.

Whilst critiques of the early "race" and children research are extensive and arise from diverse bodies of scholarship, it is important to acknowledge the early research and what grew from it in context. The early children and "race" research grew from a desire amongst psychologists to challenge the effects and growth of racism within the United States. As Lal (2002) explains of Clark and Clark (1939), two of the most influential early researchers:

> During the 1930s and 1940s, social psychologists became increasingly well-known among progressives battling race prejudice. By the early 1950s, African American psychologist Kenneth Bancroft Clark had become deeply involved with the National Association for the Advancement of Colored People's battle against segregated education in the South designed to counteract the lasting social and psychological effects of American racism. By this time, his wife, who is less well-known in the annals of history, was developing her own reputation as the guiding spirit behind Harlem's Northside Center for Child Development.
>
> Although Mamie Clark was one of the first female African American psychologists of her time, her achievements cannot be understood without knowledge of numerous historical trends, including the increasing interest in the psychology of race, and the persistence of gender discrimination in the profession. (p. 20)

Remembering and acknowledging the spirit of humanity and determination to challenge "racial" injustices that has driven much of the children and "race" research is important, even as we engage with the critique of it. The work of the early children and "race" researchers and much that has followed has played a powerful and significant role in challenging discourses of children's "racial" innocence and building a knowledge base that informs antiracist educational practices.

Diversifying Methods to Explore the Politics of Daily Lives

Partly, in response to critiques of the children and "race" research, there is a growing body of ethnographic studies of children and "race" in which researchers observe the daily dynamics of children's relationships in situ over time, rather than in a predetermined research setting (e.g., Connolly, 2000; Moore, 2002; Skattebol, 2005). Ethnographic researchers observe and sometimes probe children's views, ideas, and experiences as they unfold in situ. In studies of "racing" in early childhood settings, their data analysis often focuses on how discourse shapes the "racing" of young children's identities and how children perform their identities. In particular, much of this research has tried to extend knowledge about children and "race" by studying their peer relationships. For instance, Connolly (2000) studied the peer relations of five- and six-year-old South Asian girls within an English multiethnic, inner city primary school to develop an understanding of how racism worked within the peer group in that very specific context. Ethnographic, qualitative research of this type is seen to have many of the strengths often lacking in one-off, attitude measurement research studies of children. One-off studies are seen by some to provide a less accurate and more artificial sense of what children think when compared with ethnographic research in which researchers study the reality of "race" enactment in real lives, in real contexts over time.

The Preschool Children's Constructions of Racial and Cultural Diversity (PCCRCD) project sits within this growing tradition of children and "race" research. It has combined ethnographic research with several methods that owe much to the early and pioneering children and "race" research. Specifically, it used the following methods to explore how young children (three to five years of age) enacted and performed their "racing" in four Australian early childhood settings:

- Self-portraiture and portraiture
- Observation—structured and unstructured

- Ethnographic feedback
- Child interviews
- Story with dolls

We will briefly outline each method, how it supported us to explore the politics of young children's "racing," and suggest how educators could adapt these methods as part of their antiracist education efforts in early childhood spaces.

Self-Portraiture and Portraiture

Self-portraiture and portraiture can be used to talk with children about how they understand their embodied similarities and differences and those of their peers. In the PCCRCD project, we asked children who participated in the study to look at her/his face in a mirror. Children were then invited to talk about their physical appearance. We asked questions such as:

- Can you tell me what color your skin is?
- What color are your eyes?
- Can you tell me what color your hair is?
- What color are your lips?
- What color is your nose?

Children were also asked to draw a portrait of their friend and were asked similar questions in this process as when they drew a self-portrait.

Children were invited to choose from a variety of skin-toned pencils and pens to draw a self-portrait and were expressly asked to choose the color that looked most like their skin color. Some of the children placed their hand next to the pencils to match the color.

Using self-portraiture and portraiture in these ways made it possible to connect with the dynamics of children's lived experiences and their negotiations of subjectivities as connected to their and others' "race" identities. These negotiations were highlighted and made tangible through asking questions of children while they chose colors and while they drew opening possibilities for discussing skin color and its connection to "race." These methods of placing "race" into account are useful and useable for early childhood educators too. Drawing is a common activity in early childhood spaces and extending these spaces to use portraiture to locate children's complex negotiations of "race" in these contexts is possible and desirable in an antiracist approach. Supplying a variety of colored paper in skin tones and skin-colored pencils and being willing to ask children questions about their choices and what they mean for them opens up dialogue with children about their perceptions and preferences around "race."

Observation

In the PCCRCD project, we used a commonplace early childhood practice of observation to locate and map the dynamics of "race" across a group of children. We developed and used an observation format that logged:

- Individual and groups of children's location in the room
- Area children were playing/working in and their game/storyline
- Characters specific children played/took
- Who led game/work and who allocated roles
- Who had an active role, who followed direction, and who were on the fringes watching

We collected observations using this format over a period of eighteen months in all four services.

This mapping of the dynamics of the group and the active reading using "race" as a consideration made it possible for us to see the hierarchies of "race" in existence in the early childhood space. Layering the observation formats across each other over time highlighted patterns of inclusion and exclusion that were based on "race". Children from nondominant racial backgrounds were actively excluded by children from the dominant racial backgrounds. What this mapping further highlighted was that this exclusion was often quiet and below the radar of the educator's gaze. Further, we could see how for young girls and young boys, although racially based exclusions were just as fiercely negotiated and enforced, the experiences and enactments of these were different.

This use of observation as a tool to place "race" into account in the daily lives and practices of young children is equally powerful and possible in the early childhood educative space. Shifting normal practices of observation to be one more of group mapping, watching for patterns of inclusions and exclusions, and mapping "race" identities across this makes it possible for the educator to begin to explore how the active negotiations of "race" work in an early childhood space. This provides ways of understanding and locating where antiracist pedagogies might focus their work.

Ethnographic Feedback

Ethnographic feedback involves layering into research the feedback from the participants or focus of study. In the case of the PCCRCD project, we actively sought feedback from children and educators about the data we were collecting. This feedback sat alongside our data in order to provide it with a contextual depth and richness. This feedback is especially important in any work that seeks to explore understandings of an issue as it provides the possibility of exploring these understandings from multiple perspectives of all involved.

We sought ethnographic feedback in a variety of ways:

- We asked educators of their understandings of the children they worked with and how they saw "race" intersecting in their lives
- We asked children if they wanted to give feedback on the observations we had taken. If they wanted to give feedback, we read out the details of the observation and asked them the following questions:
  - Did they have anything about the episode they would like to tell us?
  - Had we documented what they thought happened?
  - Was there something about this episode they would like us to know?
- We asked children if they had anything further to tell us or would like us to know after and within each interview and within the small group time.

This process of ethnographic feedback provided us with further possibilities for seeing how "race" was negotiated and enacted in the early childhood spaces we were in and provided all participants with the opportunity to fill us in or elaborate on things we had missed and/or overlooked. We found that children took this process seriously and it was often in these feedback moments that children spoke of patterns of repeated racially based exclusions. Actively using this feedback approach in antibias education practices would make it possible for educators to locate exclusions and negotiations of "race" that are quiet and pass by unnoticed. In our research experience, we found that once children noticed that we employed this practice regularly and we noted down and listened to what they had to say, they were generous and open about their experiences. As an educator's tool, using ethnographic feedback processes would support the locating and challenging of "race" in the active negotiation of children's subjectivities.

Child Interviews

The opportunity for children to speak about "race" one-to-one within the PCCRCD project proved invaluable for exploring the intersections and constructions of "race" in young children's lives. In the PCCRCD project children participated in two interviews:

- One with the Persona Dolls in which they talked about their own and others differences and similarities to the Dolls, their Doll preferences and knowledge of other cultures
- One with the process of self-portraiture and portraiture

In these interviews, children were asked to name themselves and others and to choose and explain why they chose particular Persona Dolls for specific activities. During these interviews it became clear that children are able to, and do, understand and use political constructions of "race"-color and do negotiate preferences based around their understandings and constructions of "race." Hence, child interviews and one-to-one discussions with children makes it possible for an educator using an antiracist approach to unravel and locate the place of "race" in children's lives and its salience in the decisions they make in the day-to-day. This knowledge can help educators to problematize and challenge inequitable negotiations and constructions of "race" in early childhood spaces.

Story with the Dolls

Another method used in the PCCRCD project was storytelling with Persona Dolls (see Mac Naughton, 2004). We developed stories based on data we had collected in each of the four early childhood settings. The stories drew from incidents of racism and exclusion that we had observed in the centers participating in the study and/or based on issues/comments some children had made. These incidents were reshaped to fit a specific Persona Doll identity relevant to the context, and the stories were used to elicit responses from the children about:

- what they thought about what had happened?
- whether they had seen something like this happen?
- what they thought the Persona Doll character could do about this?
- what they could do?

We told the stories in small groups of four to five children and found the stories worked to engage with children's subjectivities and lived experiences of inclusion and inclusion in the daily lives of their early childhood spaces. Many children involved in the storytelling responded by sharing episodes of unfairness they had seen and witnessed. We also found that while children knew of racism occurring, they were also willing, confident, and creative in their approaches to challenging this racism.

As an educative tool, using Persona Dolls and storytelling to name racism and discrimination can actively bring racism into a public space, enabling children to give voice to the lived experiences, unfairness, questions and frustrations they experience around "race." Using this as a technique in an antiracist education also enables children to raise matters and issues of "race," knowing that the adult is willing to talk about racism with them and acknowledge "race" as a salient factor in their lives. This supports problematizing and challenging racism when it arises on other occasions, as children learn that an educator thinks it's okay and important to talk about "racial" injustice.

We now turn to the data generated in the PCCRCD through these different methods to explore how young children enact their daily "racing." In chapter 3 we begin with the role that self-portraiture and portraiture played in this work.

CHAPTER 3

The Dynamics of Whiteness: Children Locating Within/Without

Karina Davis, Glenda Mac Naughton, and Kylie Smith

Whiteness and Its Politics

In this chapter we explore how discourses of whiteness construct and reconstruct diverse identities and possibilities for children in their worlds. We ask, what does this mean for early childhood? Do we need to rethink how we work with young children around the politics of "race" in order to create greater equity for all? As Foucault (1985) argued:

> There are times in life when the question of knowing if one can think differently than one thinks, and perceive differently than one sees, is absolutely necessary if one is to go on looking and reflecting at all. (p. 8)

Now is an especially important time to find ways to know the politics of "race" differently. Debate about immigration, especially Asian immigration, has been a constant feature of the political landscape (Hage, 1998) in Australia and other Western-industrialized societies, but in our post-9/11 and post-Bali, London, Glasgow-bombings world, racist discourses are growing and hardening to normalize white and to demonize "otherness" in many countries internationally. There is a growing "Islamaphobia" (see Van Driel, 2004) in many Western countries and increasing ethnic tensions and divisions internationally. Against this background we seek in this book to rethink how we know "race" in young children's lives by focusing in this chapter on the place of whiteness in young children's "racial" knowings.

As we discussed in chapter 1, eighty years of international research on children and "race" has identified that the preschool years (three to five years of age) are critical in the development of cultural and "racial" identity and attitudes (Aboud and Doyle, 1996; Hirschfeld, 1995; Johnson, 1992; Ramsey, 1991). However, as we illustrated in chapter 1 this research on children's racial identities has been primarily cognitively oriented, focused on measuring children's racial attitudes to assess their identity development, and has presented the racialization of children's identities using developmental stage-based models (Mac Naughton and Davis, 2001).

We argue throughout this book that the racialization of children's identities is a political strategy that produces and reproduces whiteness as a position of privilege as it demonizes otherness. We also argue that politics of discourses of white identities and their effects on young children's identities vary according to children's gender, class, and ethnic identities. In this chapter, we introduce rhizoanalysis as a poststructuralist tool of analysis that educators can use to see and to trace how the politics of discourses of white identities work in forming the identities of young children whom they educate. Educators can use the processes of rhizoanalysis to generate knowledge that can deepen their antiracist work with young children.

Rhizoanalysis as a Tool of Analysis

Rhizoanalytic thought—like other poststructuralist thought—eschews the existence of a "neutral system of representation" (Mansfield, 2000, p. 143). It uses intertextuality (the way in which one text links to another text) tactically to imagine new ways to produce meaning. Rhizoanalysis is a process for exploring what a text (e.g., a research moment) "does, and how it connects with other things (including its reader, its author, its literary and nonliterary context)" (Grosz, 1994, p. 199). A rhizoanalysis aims to produce new meanings by using a tactically chosen text to "caste a shadow" over another text and, by doing so, to disrupt and challenge the politics of the initial text. It is a helpful tool for exploring the politics of identity in young children's being and becoming and can be used by educators as a pedagogical tool to engage with and disrupt unjust politics of "race."

Deleuze and Guattari (1987) suggest that a rhizomatic analysis draws on six principles: connection, heterogeneity, multiplicity, asignifying rupture, cartography, and decalcomania (Deleuze and Guattari, 1987; Hamman, 1996). In this chapter, we draw on Gilles Deleuze and Felix Guattari's (1987) idea of rhizomatic structures to explore the dynamics of whiteness in young children's lives. We combine the idea of rhizomatic structures with multiple theoretical perspectives including poststructural and postcolonial theories, to explore how young children's constructions of their racial identities are related to whiteness and to raise further pedagogical questions about how best to engage with the racialization of young children's identities in and through early years pedagogies. Educators can employ these same strategies to complexify and deepen their understandings of how young children's identities are being "raced" in their specific contexts.

A Cartography of Whiteness: Locating Whiteness and its Diverse Effects

To find political linkages and entry points into children's racialized ways of knowing, we need to locate whiteness and its diverse effects in children's lives by examining the social practices and discourses from which children draw as they negotiate links between "race," culture, and identities. Whiteness discourses link intimately with discriminatory social structures. They create the illusion that white individuals and communities can choose whether and how to engage with issues of "race," discrimination, and citizenship. This, combined with white practices of privileging the individual and their actions and effort, allows white people to ignore or deny their part in individual, structural, and institutional discrimination. Whiteness maintains its power through being "unseen" (Holmes, 2005, p. 175). Mapping whiteness makes it seen and therefore it can be used to help educators to begin to challenge the power of white in the lives of the young children with which they work.

We begin our cartography (mapping) of whiteness by locating a moment of whiteness in the life of an Anglo-Australian girl in the PCCRCD project (see appendix)—Spot (NB: a pseudonym chosen by the child). This moment occurred in a conversation with Spot about the similarities between particular dolls and people that the children knew. This conversation is our entry point to "knowing" the racialization of young children afresh through mapping the dynamics of how they can locate within and without whiteness.

Locating within Whiteness: White Skin Like Me (Spot)

Karina: Can I ask you which of these dolls looks most like your friend?

Spot: That one.

Karina: Franca. Franca does?

Spot: Yes.

Karina: And what about Franca looks like your friend?

Spot: . . . Ahh . . . Because, 'cause I think she's the prettiest.

Karina: You think Franca's the prettiest? What about Franca makes her look pretty? I'd like to know.

Spot: Ahh, because she has white socks and I like white and she has blue jeans and I like blue and she has a green top and I like green. And she has, and she has white skin and I like white skin. And I like her hair.

In this discussion, Spot shows that she knows white in very specific ways. For Spot white skin is something likeable. She locates herself within a discourse of whiteness as desirable. To understand more deeply the effects of discourses of whiteness in her life it is important to understand how whiteness works as a discourse of power. For Deleuze and Guattari (1987), as for Foucault (1980), meanings connect to each other, to systems of knowledge, and to the production, organization, and contestation of power:

> A rhizome ceaselessly establishes connections between semiotic chains, organizations of power, and circumstances relative to the arts, sciences, and social struggle. (Deleuze and Guattari, 1987, p. 7)

The connections between meaning, power, and social struggle are shifting and complex, because each point of a rhizomatic system is connected nonhierarchically to another. We can't identify whether one meaning came before or after another, only that particular meanings connect to produce and organize relations of power (Deleuze and Guattari, 1987; Hamman, 1996). Meanings are permeable to other meanings and cross into each other from "apparently separate domains" (Mansfield, 2000, p. 144). Whiteness is a discourse that continues to produce and organize relations of "race" power. It is political. It works to do this in different ways in different contexts but mapping whiteness involves mapping the differences in power and privilege between "white" people and "nonwhite" people and the advantages gained by being white in a specific context. Often these advantages are taken for granted and not talked about. Advantages can include not experiencing "racial" discrimination, not having to think about "skin" color, and being the norm against which "others" are judged.

Seeking Heterogeneous, Semiotic Connections to Power: Chains of Whiteness, Power, and Social Struggle

To analyze the moments of whiteness in children's lives rhizomatically, we must identify how meanings of whiteness link in chains of meaning to the organization of racialized power and struggle in apparently separate domains. We can do this by seeking chains of meaning between children's text and texts that explore how "race-power" is organized and struggled over in non-early childhood domains. For instance, seeking chains of meaning between children's texts and civil rights texts, autobiographical texts of those who have struggled against "race-power," or cultural studies texts of how "race-power" is organized and struggled over in the arts or sciences.

It is with antiracist political intent that we need to choose the texts we use to explore how "race-power" is organized and struggled over because rhizoanalysis aims to elucidate the operation of power. We would add that our approach to rhizoanalysis also aims to elucidate the operation of power in order to challenge oppression and discrimination and to expand human possibilities for social justice. In this instance, this means that we must choose texts that support an antiracist political intent. To do otherwise would be highly problematic for our work. For instance, if we choose to establish chains of meaning using texts with fascist intent, we would inevitably remap fascism, thus rewriting racist understandings of "white" as an unexamined and masked position and institution of power and privilege. The intent in our rhizoanalysis of the dynamics of whiteness in young children's lives is to map the racism that discourses of white identities produce in order to make it seen. Making whiteness seen disrupts the position of privilege that accrues to whiteness in contemporary Western societies.

Employing Decalcomania to Generate Multiplicity: Disrupting Discourses of Privilege

We can disrupt discourses of privilege (for instance, whiteness) in one domain (for example, early childhood education) by using texts from other domains (for example, autobiography) as decals. A decal "is a graphic device that acts as a transfer"; and decalcomania is the process of using such a transfer to "fix" a design to something. Fixing the discourses of a text from one domain to another transfers the text's meaning/s, offering us potentially new or unknown political links and entry points into how texts work.

To choose our decalcomatic texts with political intent in ex-British colonies in countries such as Australia, New Zealand, South Africa, Canada, and United States with their histories of colonialism and whiteness, we need to choose texts that theorize whiteness and post-colonialism. Such texts offer entry points to the semiotic chains that organize and contest racialized power for Anglo children as they seek to situate, explore, and deconstruct the development, existence, and maintenance of power and privilege in white, ex-colonial countries across diverse social and political domains.

If we layer or juxtapose against each other texts that do and don't produce privileged discourses of "race" in early childhood, we can disrupt the dominant discourse by demonstrating that other meanings are possible (Alvermann, 2000). The dominant discourse begins to lose its authority. Multiplicity grows by unceasingly seeking alternative decalcomatic texts that produce new "lines of flight" and, in doing so, offer new sites at which the decal can be deployed. Lines of flight resist the dominant and privileged. They are "tangential catapults that fling us out of the spiral of domination" (Fleming, 2002, p. 202).

Practicing a Rhizoanalysis of "Whiteness" and the Child

In what follows, we use the principle of decalcomania to select texts with political intent and place them over our research text to produce nondevelopmental lines of flight (Deleuze and Guattari, 1987, p. 21) about the meaning/s of whiteness in young children's identities. This process is necessarily unfinished—we could create multiple connections and possibilities of disrupting discourses of the racialized child as a developmental phenomenon, but we draw from only four decal texts in this chapter.

We chose as our decal texts voices that are persistently absent from Australian early childhood texts and that speak to specific political struggles over "race" in Australia. In other countries, different texts are likely to be appropriate but the principle is the same. It is important to choose texts that are not present in the early childhood education texts of the specific context in which you work. In choosing three of these texts, we attempted to shift the "gaze" of the "other" onto us as white Anglo-Australian meaning-makers, heeding the words of Aileen Moreton-Robinson, a Geonpul woman and scholar from the Quandamooka (Moreton Bay) area of Australia:

> In academia, it is rarely considered that Indigenous people are extremely knowledgeable about whites and whiteness. It is white scholars who have long been positioned as the leading investigators of the lives, values and abilities of Indigenous people. Indigenous scholars are usually cast as native informants who provide "experience" as opposed to knowledge about being Indigenous or white. The knowledges we have developed are often dismissed as being implausible, subjective and lacking in epistemological integrity. This is despite the fact that colonial experiences have meant that Indigenous people have been among the nation's most conscientious students of whiteness and racialisation...Indigenous knowledge of whiteness is more than a denial of dominant assumptions regarding the reality of "race" and the superiority of whites; such knowledge is not simply a reaction to what whites do and say. Our curiosity, compassion and knowledge of what constitutes humanity inform our consideration of a variety of white behaviours, histories, cultural practices and texts. (Moreton-Robinson, 2004, p. 85)

Decal Text 1: Aunty Iris's Text

Creating Historical Lines of Flight to Identify Racialized Dichotomies (Black/White) as a Performance of White Sovereignty

> They were taken because they never lived like white people and the Government wanted to turn them into white kids.

Aunty Iris's text describes her lived experiences as an Indigenous Australian woman growing up in 1920s and 1930s Australia. Her text creates a line of flight from Spot's text where white is "liked." It allows us to consider how the racialized identity of whiteness as desirable was a politically constructed brutality. This brutality was produced in and through the removal of Indigenous children from their families and their placement with the desirable "white" Australian families or institutions.

Against this line of flight, how might we position Spot's text "And she has, and she has white skin and I like white skin"? Is it a four-year-old child's apolitical color categorization of the world? Is it an instance of white sovereignty performance (Nicoll, 2004) in which white

Australians exercise power over Indigenous identity by continually reinscribing the historically racialized dichotomy of "white/black" as "desirable/undesirable"? Is Spot performing white sovereignty? Can young children perform white sovereignty?

Decal Text 2: Ien Ang's Text

Creating Academic Lines of Flight to Identify Developmental Discourses of "Race" as a White Imperialist Strategy

> It is important to emphasize, at this point, that white/Western hegemony is not a random psychological aberration but the systematic consequences of a global historical development over the last 500 years. (Ang, 2003, p. 199)

> It is this historical sense that the hierarchical binary divide between white/non-white and Western/non-Western should be taken account of as a master-grid framing the potentialities of, and settings limited to, all subjectivities and all struggles. (Ang, 2003, p. 199)

Ien Ang identifies herself as an Australian woman of Chinese descent. Her text creates a line of flight that leads us to ask, "Is the global historical development of child development discourses a systematic effort to maintain a hierarchical binary divide between white/nonwhite ways of knowing 'race'?" As an allegedly universal and apolitical paradigm, developmentalism explains children's racist speech and actions in terms of their lack of cognitive capacity to understand its meaning, arguing that children aged three and four do not understand issues of color, "race," privilege, inclusion, and exclusion. In contrast, whiteness theories regard all speech as strategies to normalize and privilege white. When Spot said, "And she has, and she has white skin and I like white skin," was she acting intentionally to mark white as desirable and normal? Is whiteness the master grid framing her subjectivities? We generate these questions to produce another possibility for how we make sense of young children's racialized speech and actions. In this other possibility whiteness is seen.

Decal Text 3: Judy Atkinson's Text

Creating Biographical Lines of Flight to Identify Whiteness as Always Better, Inevitably Desirable?

Judy Atkinson is an Indigenous Australian academic woman who has drawn on biographical stories of Indigenous women and men to begin the process of "healing" the generational trauma resulting from the racist and violent treatment of Indigenous communities. Lorna's is such a story. Lorna describes her childhood as a political struggle against a masculine white hegemony that positioned her as inferior and undesirable because of her "black" identity:

> I could never be as good as the white man that was in charge that was the overall feeling of my childhood. (cited in Atkinson, 2002, p. 99)

Against this line of flight how do we understand Spot's text, "And she has, and she has white skin and I like white skin."? Is Spot's preference for "white" just a natural preference for her own group? Will she ever feel that "white" is never as good as black? Is her capacity to like "white" historically and politically inevitable? Is naming "white" in Australia always naming a relation of power and desirability? Is it always political?

Expanding Our Cartography of Whiteness

These questions become more challenging to explore when we sit them against other PCCRCD moments of whiteness in which Chinese-Australian children chose to locate themselves within discourses of white as desirable. Two such moments occurred during two interviews Kylie conducted with Kung Fu (his chosen pseudonym) a Chinese-Australian boy who was four years old at the time we interviewed him. In the first interview Kung Fu spoke with Kylie about himself and drew how he looked. In this interview, Kylie asked Kung Fu, "Can you tell me what color is your skin?." He replied, "White." Kylie then asked him, "Would you like to choose a colored pencil that matches your skin color so that you can draw yourself?" Kung Fu chose a melon-colored pencil, one of the lightest skin-toned colors in the pack.

When Kylie asked him in the second interview to tell her who his friends were and what their skin color was, he said that his friends Green Goblin, Snake, Spider, and Superman were each white. Like Kung Fu, each of his friends was a Chinese-Australian boy. In these moments of whiteness Kung Fu located himself and his friends firmly within whiteness.

In what follows we transpose the questions we generated about Spot to Kung Fu to extend our exploration of whiteness in young children's lives. Our transposition generates questions such as, "Is Kung Fu's preference for 'white' just a natural preference to belong to the dominant group?" "How does this intersect with the power of whiteness in both and across Australian and Chinese contexts?" "Is his capacity to like 'white' historically and politically inevitable?" "Is naming 'white' in Australia always naming a relation of power and desirability?" and "Is naming a relation to white always political?"

We now return to our rhizomatic principles of seeking chains of meaning that connect to the organization of power to explore these questions further. Our intention this time is to create connections between Kung Fu's identity texts and current hegemonic discourses of "white" in Australian society to decenter it and create a point of rupture in its dominance. We "flee" from assuming that his interview text is apolitical and draw new lines of flight to consider its politics. We begin this process by choosing as our fourth decal text, a text from a journal about human rights in Australia that links whiteness to immigration.

Decal Text 4: The Australian Journal of Human Rights

Creating Contemporary Lines of Flight to Identify Naming Whiteness as Naming the Right to Australian Citizenship

> Australia's borders have historically been racialized and premised on whiteness. They have provided the parameters for the inclusion and exclusion of certain peoples along racialized line. (Tascon, 2002, p. 3)

Tascon's decal text highlights that in contemporary Australia being "white" determines who is seen as a "real" Australian citizen and who is not. Of the children in Kung Fu's center, 92 percent were from Chinese-Australian or Vietnamese-Australian backgrounds. When the families in Kung Fu's center were initially invited to participate in PRCRD project, many were hesitant to do so because of the need to divulge their home address on the child consent form. Several of those parents linked their hesitancy to their status as immigrants in Australia. The parents' concerns about their immigrant status and Kung Fu's efforts to connect to white arose at a time when contemporary Australian government discourses about Australian citizenship were actively and deliberately reinscribing whiteness as a nonspoken yet essential ingredient to be a "real" Australian citizen. The "parameters for the inclusion and exclusion of certain peoples along racialized lines" (Tascon, 2002, p. 3) was hardening.

It is difficult to judge how these exclusionary, racialized discourses of citizenship touched Kung Fu's life through his parents' concerns or through other mechanisms. However, by creating lines of flight from Kung Fu's research text to texts of human rights, "race," and immigration in Australia, we can produce a semiotic chain connecting Kung Fu's research text with exclusionary discourses of white identities, citizenship, and immigration that can normalize whiteness in the lives of young children and restrict the constructions that Kung Fu might imagine or desire for himself and his friends as immigrant "others" in contemporary Australia. This chain of meaning links to the organized power of whiteness in contemporary Australia and is strengthened if we further draw on Tascon (2002) as a decal text. This creates contemporary lines of flight in which naming whiteness is seen as connected with the right to Australian citizenship.

> Onshore refugees, "boat people," intersect with the Australian border in its manifold manifestations in a most violent manner. We cannot name it an interaction, as that would imbue the contact with an equality it does not possess. These refugees are some of the most vulnerable people on the globe, needing to leave the safety of their own borders to cross others', to enter spaces possibly hostile to them. Australia's space has been hostile to them. (Tascon, 2002, p. 7)

Increasingly, acts such as the increased aggressive patrol of the Australian borders and the introduction of English language proficiency testing for immigrants are whitening the discourse of Australian citizenship and producing an exclusionary discourse of citizenship, with "citizenship privilege" (Choules, 2006, p. 276) accruing to white Australians. Increasingly, Australia is a "hostile" (Tacson, 2002, p. 7) space in which to be other than white. Was Kung Fu marking himself and his friends as "white" because he had already normalized and assimilated knowledge about how hierarchies of "race" and citizenship operate in contemporary Australian society? Or was Kung Fu marking himself and his friends as white in response to hierarchies of "race" and class in contemporary Chinese contexts? How do these intersect for Kung Fu with his experiences as an immigrant in Australia?

However, whiteness is not solely an Australian invention. The power and privilege that being "white" brings is also established through traditional Chinese caste systems in which darker skin often signifies a person is of a lower-caste background (Dikotter, 1994; Kibria, 2000). A person of a lower-caste background is likely to have limited or no education, lack employment, and lack the capacity to purchase consumer commodities. A lighter skin is more highly valued and identified as being higher caste (Taylor, 2004). As the author of Skin Deep explains:

> Skin whitening has a long history in Asia, stemming back to ancient China and Japan, where the saying "one white covers up three ugliness" was passed through the generations.... A white complexion was seen as noble and aristocratic, especially in Southeast Asia, where the sun was always out. Only those rich enough could afford to stay indoors, while peasants baked in the rice fields. (Skin Deep: Dying to Be White [CNN])

Perhaps, given the intersections of these differing, yet powerfully intersecting discourses of whiteness in Kung Fu's life, it is not surprising that he might seek for himself and his friends what Bourdieu (1990) calls social, cultural, and symbolic capital that whiteness could bring a person living in Australia who is a Chinese-Australian. Whilst rhizoanalysis cannot establish as fact that Kung Fu desired whiteness or that he was attempting to accumulate its benefits, it does ask us to consider the possibilities.

From Lines of Flight to Rupturing Strategies of Whiteness: Toward Alternative Pedagogies

Cognitively oriented child and "race" researchers have explained children's racial identification as "merely" an apolitical, developmental phenomenon in which children build a positive affinity with their cultural identities by noticing and naming physical difference. As we have discussed in earlier chapters, pedagogically this is highly problematic. It leaves the educator waiting for developmental phenomenon to unfold, and it denies the complex political learning that produces it.

In contrast, in this chapter, we have used rhizoanalysis of young children's texts with overt political intent to generate other questions about the phenomenon of young children's racialization. Those questions include:

- Can young children perform white sovereignty?
- Is Spot learning how to name her structurally privileged position in Australian society and is Kung Fu learning how to accumulate that privilege for himself and his friends?

- Is Spot's preference to like "white" historically and politically inevitable? Is naming "white" in Australia always naming a relation of power and desirability?

These questions make white discursive social practices and negotiations of identity a potential focus for antibias, antiracist pedagogies that challenge white privilege and supremacy as a source for identity building. They challenge us to think about the pedagogical necessity and possibilities of using the tactics of rhizoanalysis to redirect teaching effort in three key ways.

First, teaching effort needs to be redirected toward locating, mapping, and challenging the strategies through which whiteness recycles power to produce whiteness as desirable and normal in the early years space. For educators, this suggests the need to develop a new language and set of strategies to drive their observation and assessment of young children's "race" learning. Instead of observing the developmental unfolding of the child, it becomes important to do a cartography (mapping) of whiteness in the space so that white is located and seen. In this work educators need to reflect on the knowledge they use to make sense of what they see. Questions that can help in this process are:

- Whose knowledge of children am I deploying to make sense of "race" in young children's lives?
- Whose knowledge is my choice privileging and whose knowledge am I silencing?
- Whose knowledge am I struggling to find and/or to use?
- Whose knowledge is it important to use given the relations of "race" and power that exist in my region, country, city, town, community?
- Who has been struggling to challenge unfair relations of "race"-power in my region, country, city, town, community? How might I use their texts as "decals" to make "race"-power seen?

Second, educators need to consider that what children say in the daily life of an early childhood space is political. For educators this means becoming familiar with the political texts of "race" and "race" struggles for social justice relevant in their region, country, city, town, and/or community. To understand that "race" is a political struggle requires a familiarity with the terrain of those struggles. Instead of relying on developmental texts to know the child, it becomes important to know the child through "race"-struggle texts. For instance, educators could draw on the many national and international autobiography and biography texts that document "race" struggles of individuals, in which they also describe their childhoods as structured by racially based inequities and injustice. This provides "other" opportunities to consider how, for many, "race" structures lives, as well as working to highlight intersections with how children can and do construct and actively negotiate the politics of "race" in their lives with understanding of its power and with political intent.

Third, teaching effort needs to be redirected toward identifying early childhood texts to use with children, which can create different lines of flight that sweep children into new possibilities for themselves and others. For educators, this means searching for texts that help children to engage with how whiteness operates as a position of privilege in their lives. Instead of seeing the racialization of identity as an innocent development process, it becomes important to directly engage with how it produces and reproduces the power of whiteness in young children's lives. For instance, Coloma (2006) describes the process of consciously and unconsciously working against the normalization and assimilation of "Asian" racialized groups as disorienting. She compartmentalizes the word into three sections to remake the concept it conveys into an active political strategy for being Otherwise. First, she separates the beginning "dis" to mark a performance to intensely negate, interrupt, disrupt, and/or contest a person's identities (Coloma, 2006). Choosing texts that can help do this is a priority in de-"racing" the early childhood curriculum. Second, she separates the term "orient" that means an "exotified, imperialized and static symbol of Asia, but also the act of alignment and positioning in relation to others" (Coloma, 2006. p. 2). Third, she separates "ing" because she argues that it "signifies something active and ongoing" (p. 20). A priority in de-"racing" the early childhood space would be to find texts that disorient Kung Fu and children who are othered within the specific discourses of whiteness in a specific early childhood space.

Final Reflections

Discourses of childhood racial innocence and color-blindness can bring concern that talking about racism is harmful to young children. Those challenging the need for early antiracist education may believe that disrupting children's innocence about racism will do more harm than good in two ways. It can create emotional distress in young children from a variety of racial backgrounds and it can increase racism by bringing children's attention it. There is some (see Kehoe, 1984), but fairly minimal, research evidence to support this view (Mansfield and Kehoe, 1994) in studies with older children and youth.

More hopeful for advocates of antiracist education is recent research by Hughes, Bigler, and Levy (2007) with primary school children in the United States and found positive effects from this teaching on children's racial thinking and feeling:

> European American children who learned about historical racism had more positive and less negative views of African Americans than did children who received similar lessons that did not include information about racism. (p. 1701)

They also found that:

> Learning about racism may not be as harmful as some individuals have predicted. Although learning about racism was associated with higher levels of racial guilt and defensiveness among some European American children, these responses may not be wholly negative. (p. 1701)

As with many oppressive systems, whiteness discourse privileges those who are dominant—that is, those who are white. It allows whites to live and function more freely by enabling whites to ignore their "race" and culture while rendering invisible its positive effects for them. This privilege is further reproduced and maintained within a discourse that values white, and therefore rewards those who are white (Frankenberg, 1993; McIntosh, 1997; Moreton-Robinson, 2000; Roediger, 1994). Taking lines of flight from developmental readings of "race" and children's identities engages us in a political analysis that is always in a process of becoming works *as it* to continually challenge racism. It extends the lenses educators can use to read the texts of daily life produced and used by children in early childhood spaces, as they negotiate the politics of their identities. Educators can use the processes of rhizoanalysis detailed in this chapter to locate the politics of children within/without whiteness and reflect on how best to deepen and broaden their antiracist pedagogies. If children have access to decal texts that challenge them to think "otherwise" about "race," it will help them learn to challenge for themselves the idea that being white is the one right, normal, and proper way to be and to do "race." They can become part of the project of "de-racing" early childhood and celebrating the complexities of being who we are, with safety, humility, and justice in mind.

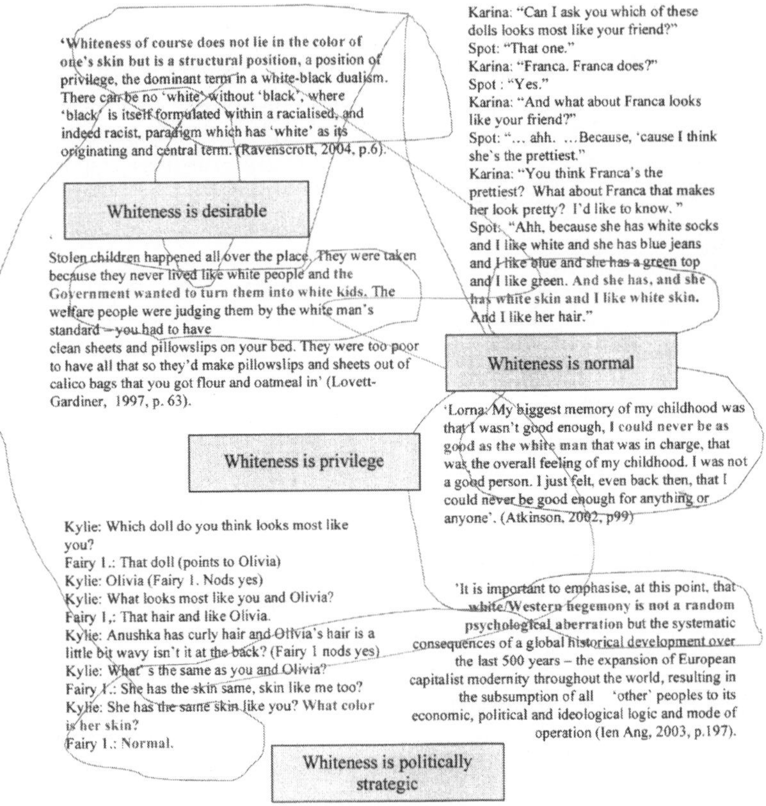

'Whiteness of course does not lie in the color of one's skin but is a structural position, a position of privilege, the dominant term in a white-black dualism. There can be no 'white' without 'black', where 'black' is itself formulated within a racialised, and indeed racist, paradigm which has 'white' as its originating and central term. (Ravenscroft, 2004, p.6).

Karina: "Can I ask you which of these dolls looks most like your friend?"
Spot: "That one."
Karina: "Franca. Franca does?"
Spot : "Yes."
Karina: "And what about Franca looks like your friend?"
Spot: "... ahh. ...Because, 'cause I think she's the prettiest."
Karina: "You think Franca's the prettiest? What about Franca that makes her look pretty? I'd like to know. "
Spot: "Ahh, because she has white socks and I like white and she has blue jeans and I like blue and she has a green top and I like green. And she has, and she has white skin and I like white skin. And I like her hair."

Whiteness is desirable

Stolen children happened all over the place. They were taken because they never lived like white people and the Government wanted to turn them into white kids. The welfare people were judging them by the white man's standard – you had to have clean sheets and pillowslips on your bed. They were too poor to have all that so they'd make pillowslips and sheets out of calico bags that you got flour and oatmeal in' (Lovett-Gardiner, 1997, p. 63).

Whiteness is normal

'Lorna: My biggest memory of my childhood was that I wasn't good enough, I could never be as good as the white man that was in charge, that was the overall feeling of my childhood. I was not a good person. I just felt, even back then, that I could never be good enough for anything or anyone'. (Atkinson, 2002, p99)

Whiteness is privilege

Kylie: Which doll do you think looks most like you?
Fairy 1.: That doll (points to Olivia)
Kylie: Olivia (Fairy 1. Nods yes)
Kylie: What looks most like you and Olivia?
Fairy 1,: That hair and like Olivia.
Kylie: Anushka has curly hair and Olivia's hair is a little bit wavy isn't it at the back? (Fairy 1 nods yes)
Kylie: What's the same as you and Olivia?
Fairy 1.: She has the skin same, skin like me too?
Kylie: She has the same skin like you? What color is her skin?
Fairy 1.: Normal.

'It is important to emphasise, at this point, that white/Western hegemony is not a random psychological aberration but the systematic consequences of a global historical development over the last 500 years – the expansion of European capitalist modernity throughout the world, resulting in the subsumption of all 'other' peoples to its economic, political and ideological logic and mode of operation (Ien Ang, 2003, p.197).

Whiteness is politically strategic

Figure 3.1 Meanings of Whiteness: Employing Decal to Produce Alternative "Lines of Flight"

Source: Mac Naughton, G., K. Smith, and K. Davis. (2005). A Rhizoanalysis of Preschool Children's Constructions of Cultural and "Racial" Diversity.

Further Resources

For those educators who would like to begin or further explore antiracist education with young children but feel unfamiliar with or unsure where to start, the following list may help your work. This list provides a number of starting points and resources for thinking about the implementation of antiracist pedagogies.

hooks, b. (1994). *Teaching to transgress: Education as the practice of freedom.* New York: Routledge.

Mac Naughton, G., and G. Williams. (2008). *Teaching techniques for young children*, 3rd ed. Melbourne: Pearson Education.

Children's Books

Coerr, E. (1993). *Sadako.* New York: Scholastic Australia.

Jolly, J. (2006). *Ali the bold heart.* Balmain, NSW: Limlight Press.

Lofthouse, L. (2007). *Ziba came on a boat.* Camberwell, Victoria: Penguin Viking.

Marsden, J. (1998). *The rabbits.* South Melbourne, Victoria: Thomas C. Lothian.

Maughan, W. (2005). *Malian's song.* Vermont, USA: The Vermont Folklife Center.

Randall, B., and M. Hogan. (2008). *Nyuntu ninti (what you should know).* Sydney: ABC Books.

CHAPTER 4

Intersecting Identities: Fantasy, Popular Culture, and Feminized "Race"-Gender

Glenda Mac Naughton, Karina Davis, and Kylie Smith

Introduction

In this chapter we explore in more depth how white discourses can work in children's lives with a focus on the politics of young girls gender-"race" identities. From the earliest studies of "race" and young children's identities (e.g., Clark and Clark, 1939; Horowitz, 1939) researchers have studied the differences between the racial preferences and awareness of boys and girls. For instance, in the Clark and Clark (1939) study they found that the sex of the child mattered to what was found:

> The most significant aspect of the results . . . is the fact that the choices of the boys show significant trends whereas those of the girls seems to approximate chance. This fact can be best understood if it is remembered that the boys were making identifications of themselves while the girls were identifying brothers, cousins, and in a few instances a boy playmate. Because of this difference in response it would appear that either the technique used in this investigation has greater validity when used with boys than when used with girls, or that the dynamics involved when girls identify someone other than themselves is quite different from the self-identification of the boys. (pp. 596–597)

In research exploring "race" and gender following Clark and Clark's (1939) study, gender has been constructed typically as a pre-given biological binary (male versus female) that can be studied as a separate yet salient variable that mediates a researcher's findings about children's development of racial awareness (e.g., Semaj, 1981; Singleton and Asher, 1979; Vora and England, 2000).

In this chapter and the next (chapter 5), we seek to rethink this construction of gender in young children's racial identities and point to the political and pedagogical implications of such a rethink. The data we use in this chapter is again taken from the PCCRCD research project conducted in Melbourne, Australia (refer appendix for project details) on how "race" was socially constructed and gendered in the lives of twenty-three three- and four-year-old girls from three different ethnic backgrounds. These girls were Vietnamese-Australian (four girls), Chinese-Australian (six girls), and Anglo-Australian (thirteen girls). In what follows we show how these girls negotiated ways of naming themselves and their friends that inextricably and intimately gendered their "racing," and "raced" their gendering of identities through their relationship to what we term "proto-feminized whiteness."

At the end of the chapter we raise questions about how we can challenge the unjust racial politics embedded in proto-feminized whiteness as part of broader antiracist early childhood pedagogies. Proto-feminized whiteness binds idealized norms of feminine beauty from within Western popular culture with the racialized privileges and strategies of whiteness to produce beauty as inseparable from whiteness. In this, it produces race and gender as inseparable from each other and each as inseparable from young children's culturally mediated constructions of identity. These culturally mediated constructions have implications for how we work in early childhood spaces in ways that are fair and equitable for all.

We explore how the girls in the PCCRCD project engaged in a "localized identity politics" of "race"/gender that embedded wider, culturally mediated, gendered, and racialized privilege in their daily lives; and continue our argument that feminist poststructuralist studies of children and "race-gender" identities can add considerably to the children and "race" research by responding to Hall's (1999) caution that:

> Discussions of race that are not informed by an awareness of the ways in which race is gendered and gender is raced are incomplete. (p. 2)

We would add that such discussions are also incomplete if they ignore the cultural mediation and making of identities that young children actively engage in and negotiate, as they engage with the politics of their cultural worlds in early childhood education and beyond.

In what follows, we draw from data from two of the early childhood centers that participated in the PCCRCD project and analyze data generated by children's choice of pseudonym, individual discussions with children, and field observations of their play. We begin our exploration of the gendering of "race" identities by using critical theories of whiteness to interrogate the pseudonym choices made by the Chinese-Australian, Vietnamese-Australian, and Anglo-Australian girls who attended Centre 1 and Centre 3.

Gender/"Race" and the Girls' Choice of Pseudonyms

Table 4.1 shows the pseudonyms that each girl chose and each girl's ethnic background (as named by their parents). The majority of the girls (65 percent) used a female fantasy popular culture icon as their

Table 4.1 Girls' Pseudonym Choices

Pseudonym	Children's Ethnic background	Centre
Barbie	Chinese-Australian	1
Barbie 2	Vietnamese-Australian	2
Barbie 3	Vietnamese-Australian	2
Barbie Princess	Anglo-Australian	1
Barbie	Anglo-Australian	1
Fairy	Chinese-Vietnamese-Australian	2
Fairy 2	Chinese-Australian	2
Fairy	Anglo-Australian	1
Fairy	Anglo-Australian	1
Princess Fairy	Anglo-Australian	1
Princess	Anglo-Australian	1
Charlie's Angel	Vietnamese-Australian	2
Charlie's Angel 2	Chinese-Australian	2
Butterfly 1	Anglo-Australian	1
Butterfly 2	Anglo-Australian	1
Ballerina	Anglo-Australian	1
Rabbit	Chinese-Australian	2
Kangaroo	Chinese-Australian	2
Tiger	Chinese-Australian	2
Child 0128–No pseudonym	Anglo-Australian	1
Lisa	Anglo-Australian	1
Jenny	Anglo-Australian	1
Child 0133–No pseudonym	Anglo-Australian	1

Source: Author's Illustration.

pseudonym. The four icons that the girls used most often were Barbie, Fairy, Princess, and Charlie's Angels; and the two most popular icons were Barbie (21 percent), and Fairy (21 percent). Five girls (21 percent) chose characters from the natural world—Butterfly (two Anglo-Australian), Rabbit (one Chinese-Australian), Kangaroo (one Chinese-Australian), and Tiger (one Chinese-Australian). Two Anglo-Australian girls chose an alternative girl's name (Jenny and Lisa). Two Anglo-Australian girls did not choose a pseudonym.

Critical white theories focus on the social and cultural mechanisms that legitimate whiteness as the norm and produce its supremacy through how it contrasts with blackness or brownness (Morrison, 1992). They also draw attention to how white is encoded as a universal marker of desirability and privilege (hooks, 2003). We argue in what follows that the girls' choice of pseudonyms showed the workings of gendered whiteness; and, within this, that their identities were bound inextricably to proto-feminized white identities. In our view, each of the popular culture icons chosen by the girls reinforced proto-feminized white identities in several ways. Each icon produced (and continues to produce) a fantasy image of femininity that is linked discursively and structurally to "whiteness as a suppressed, invisible privilege" (Thompson, 1999). Further, it produced a desirable physical appearance as a normalized embodied feminine beauty that is seductively white and heterosexual. Beauty was gender-"race"d. To build our argument, we first focus on the mechanisms through which the three specific popular culture icons that dominated the girls' choices of pseudonyms (Barbie, Princesses, and Fairies) work to (re)-produce proto-feminized whiteness.

Popular Culture Icons of Proto-feminized Whiteness

Barbie has long been recognized and critiqued as a contemporary sign of idealized embodied feminine beauty (Lord, 1994; McDonough, 1999) and within this for her whiteness (see Chin, 2001; Ingraham, 2004; Magee, 2005). Ironically, Barbie's contemporary complicity in establishing idealized embodied white beauty as a norm arises from Mattel's efforts to deflect the extensive critical debate that has been generated about her whiteness. Mattel's response to this debate was to produce a series of "ethnic" Barbies of the world (e.g., Jamaican Barbie, Polynesian Barbie, Indian Barbie, Native American Barbie). However, this initiative has merely re-encoded white Barbie as the physical norm from which other Barbies should be measured and defined (du Cille, 1996). This has successfully produced the normativity embedded in the idealized white Barbie as more desirable and authentic than the exotic "others." Within this, the "archetypal white American beauty" (du Cille, 1996, p. 553) Barbie mold (the proto-Barbie) was used as the physical mold for the ethnic Barbies, causing du Cille (1996) to argue that the ethnic Barbies do not represent "the triumph of difference but rather that of similarity, a mediated text that no matter what its dye job ultimately must be readable as white" (pp. 550–551). So, even though "other" Barbies were produced to highlight their differences from the white proto-Barbie, she strips and merges their cultures and images with her own, denying their difference from her. This reinforces that she is the authentic Barbie from which all others are derived and from which they must be referenced.

It is not only proto-Barbie's physical features that privilege idealized norms of white beauty by molding other ethnicities to them. In giving each ethnic Barbie a specific culture and geographical location, Mattel make proto-Barbie universally located both culturally and geographically. Proto-Barbie is not "Jamaican," "Polynesian," "Indian," or "Native," she is just "Barbie," citizen of everywhere and belonging to and with everyone's culture. Like whiteness itself she does not need locating geographically or naming culturally because she is the norm to which all "othered" Barbies defer and refer. She is the supreme universal against which all others must be located and named. This supreme universal is white and North American, but this is never said. As Varney (1998) argued:

> Barbie simply collapses all identities into her own, robbing children of the breadth of cultural links and understanding of their own cultures and appreciation of other cultures which, arguably, should be open to them. (p. 161)

Nowhere is this collapsing of identities more evident powerful than in the positioning of "ethnic" Barbies against proto-Barbie. Locating and positioning her as proto-Barbie effectively collapses all gendered identities into whiteness, and as such produces whiteness as the desirable and privileged position to which all girls should aspire. This re-encoding of proto-Barbie as the real Barbie is also reinforced strongly by the myriad licensed commodities (including clothing, food, musical instruments, phones, make-up, and toys) carrying the Barbie trademark. Proto-Barbie is the only Barbie whose image consistently appears on those licensed commodities (e.g., B-School writing tablet,

Barbie shampoo, Barbie toothbrush, and Barbie bike). This reinforces the idea that she is the supreme Barbie.

More specifically, Thoma (2007) draws on Yoon (1994 cited in Thoma) to link proto-Barbie's cultural and racial invisibility directly to her capacity to silence the possibilities for an Asian ethnic or racial identity:

> While white Barbie has a racial and a cultural heritage, these aspects of her dominant culture identity are made invisible in the "All-American" world that Mattel deploys. Yoon reveals that white Barbie not only sells a gender identity but also a nationality that precludes any Asian ethnic or racial identity. To be Malaysian or Chinese, East Indian or Korean, one must be from and identify with another part of the world. (Accessed online at http://social.chass.ncsu.edu/wyrick/DEBCLASS/barbie.htm#note34)

So, whilst the official Barbie website (http://www.barbie.com), might call girls who enter it to "B who U wanna B," and asks of them, "What kind of Barbie girl are you?" its text, images, and product lines clearly show that proto-Barbie is the reference point against which you should "B." Whatever you want to "B," within the Barbie collection, being "Asian" is not an option for children. As Wen points out:

> In the Barbie world, Asian-looking dolls are confined to the adult-oriented collector edition with an international focus or the second-tier "Friend of Barbie" collection. (Wen, 2000, accessed online)

As contemporary popular culture icons, Princesses and Fairies encode whiteness in equally powerful ways and use similar processes to those that establish proto-Barbie as desirable and ethnic Barbies as other. At times, within these processes they draw on Barbie or the multinational Disney brand to further reinforce the desirability and idealization of white feminine beauty. Proto-Barbie's position of white supremacy is further embedded in the Barbie Fairytale Princesses product line, where proto-Barbie becomes a princess in the fairy tales of Swan Lake, Rapunzel, Nutcracker, and The Princess and the Pauper. She also reinvents herself as a princess in The 12 Dancing Princesses. Becoming the princess strengthens her identity as the Barbie who is special, desirable, and beautiful.

Barbie was not the only contemporary cultural guide to being a Princess that the girls in our study could access. As we collected our data, Disney launched its Disney Princess line drawn from eight fairy tales:

- Cinderella: white/blond, "gentle and soft-spoken," "a keen, intelligent sense of humor," "true dignity."
- Princess Jasmine: "an exotic, fiery beauty,...doesn't want much—just to marry for love and to experience life outside the palace."
- Mulan: a princess from a Chinese folktale with no "About me" information.
- Ariel: a redheaded white mermaid—is "feisty" and a "bit naïve."
- Snow White: "beautiful" with a "pure, loveable nature."
- Aurora: from Sleeping Beauty, white and blonde, "gentle, loving and thoughtful."
- Belle: from Beauty and the Beast, "lovely as her name implies" with "natural" beauty and "inner strength."
- Pocahontas: "beautiful" "playful, free-spirited" with a "passionate spirit" and a "forest home."

(Accessed online at About me, Disney Princess official Web site: http://disney.go.com/princess/ on November 21, 2006)

Aimed at three- to six-year-old girls, Disney's Princesses are an extremely successful product line, netting Disney well over US$2 billion in sales internationally during 2004 (Petrecca, 2005, sourced online). It has done this through linking older Disney characters strategically with other fairy tale figures. As Petrecca (2005) explained:

> By linking older characters such as Cinderella and Snow White under the princess umbrella with newer figures such as Beauty and the Beast's Belle and Aladdin's Jasmine, Disney created a film, clothing and toy empire that brought in $2 billion in fiscal 2004 sales.

In linking new Princess characters with older fairytale Disney characters, the Disney brand presents and embodies beauty in similar ways to Mattel's proto-Barbie. In this case, female characters who are princesses, are known for and through their beauty, and that beauty is strongly and explicitly associated with being white, economically privileged, and virtuous (Baker-Sperry and Grauerholz, 2003, p. 724). Moreover, the female characters that appear in Disney's contemporary movies and books make strong and continual reference to the characters' physical appearance and beauty, and the movies and books that do this are the ones that are reproduced most often (Baker-Sperry and Grauerholz, 2003).

As with Mattel's Barbie product range, the Disney collection of princesses also includes ethnic "others." However, these "others" are discussed and positioned against the beauty of the white, demure (possibly "real" or true) Disney Princess. In the Disney Princess range, Mulan is a princess from a Chinese folktale, Pocahontas is a native American princess, and Jasmine is an "Arabic" princess. Each of these ethnic "others" have shorter descriptors in the "About me" section of the official Disney Princess Web site than the Princesses who are white; all are positioned as "other" within their descriptions and there is little if no mention of their "beauty." Mulan has no "About Me" information and as such is effectively silenced. Princess Jasmine is named through her otherness to the demure characters like Cinderella and Snow White; Jasmine, "exotic and fiery," is the only princess who seems dissatisfied with her role in the palace and restless with the demure role a princess needs to adopt. Pocahontas is described as being a "free spirit" and lives in her "forest home," reinforcing the stereotyped and essentialized image of Indigenous peoples and cultures as living close to nature (Kaomea, 2000; Mac Naughton and Davis, 2001).

The language used in these descriptors stands in stark contrast to the descriptions of the "white" Princesses and these descriptors raise many issues for antiracist work, as they essentialize stereotyped characteristics of others while reinforcing the ideal white feminized prototypes of beauty young girls access. When ethnicities perceived as nonwhite are connected as strongly to otherness as is done within Disney's Princesses, questions about the extent to which the ethnic "other" princesses are genuine and/or authentic invariably result. This works to reinforce the understandings that to be a real or authentic beauty is to be white and feminine in very particular ways.

Disney's Fairies product line also works in similar ways and builds from proto-Fairy images to encode highly feminized and white images of beauty. While this image of the female white fairy as beauty is not new (i.e., Sugar Plum Fairy), Disney's move to create a fairy product range that has produced a Disney Fairy brand name narrows the options for young children, especially girls, to envision and embody themselves as something or someone outside of this image. A visit to Disney's official Fairies Web site (http://disney.go.com/fairies/) shows their lead Fairy—Tinkerbell—to be white and Barbie-like with "barely black" friends. Similarly, in the chain store Target's Disney Fairies product range (backpacks, bed lights, bath towels, bedding, watches, and jewelry), Tinkerbell is reinforced as proto-Fairy. As with proto-Barbie, it is Tinkerbell's image that adorns the product range rather than that of any of her friends, and it is Tinkerbell's image that casts the image mold for others. Using similar strategies and processes to those embedded in proto-Barbie and Disney's Princesses, Disney Fairies embody idealized feminine beauty in ways that make it essential for female popular culture icons to be white to be beautiful.

The Girls' Popular Culture Icon Choices and the (Re)production of Proto-feminized Whiteness

Intersecting with these popular culture icons of Barbie, Princesses, and Fairies are the girls in this research project and their choice of pseudonyms. Whilst the girls' choice of pseudonyms may seem data that is relatively minor, even mundane, we agree with Hanks (1996) that it is through "the most mundane utterances" (p. 237) that we produce ourselves in reference to others and thus come to realize ourselves. Increasingly, young children's reference points for their sense of what is physically attractive are the "mundane utterances" of the scripts of popular culture and the media and entertainment industries (Dohnt and Tiggemann, 2006). In our study, the girls demonstrated a strong awareness of the popular culture icons of Barbie, Princesses, and Fairies and their associated scripts, and chose actively to use them for their pseudonyms. Their "mundane" choices of pseudonym connected their gender-"race"d research identities seamlessly with a proto-feminized whiteness that suffused their choices about how to represent themselves and their friends in their self-portraits and in their daily play with each other. The Anglo-Australian girls and the Vietnamese-Australian girls made choices that were similar, yet in these choices the embodiment of proto-feminized whiteness worked in complex ways that both intersected and diverged.

Seamless Connections: Anglo-Australian Girls, Barbie Pink, Princesses, and Whiteness

The majority of the Anglo-Australian girls in Centre 1 were extremely familiar with the Disney and Mattel iconography of Princesses and Barbie. For instance, several of the girls owned branded backpacks with images of Princesses or Barbies on them. In the following exchange, Child 0133 and Butterfly 2 talk about what princesses look like and what they wear, as they each drew a picture of a girl who becomes a princess. The starting point for this transition from "girl" to "princess" is a "Barbie pink" pencil:

Child 0133: I'm going to draw a girl.
Butterfly 2: I'm going to draw a girl, too.
Child 0133: Pink...pink. Where's the pink pencil? I need it for the girl.
Butterfly 2: Here.
Butterfly 2 gives Child 0133 a pink pencil that is bright "Barbie pink." Butterfly 2 and Child 0133 drew their pictures using this pencil.
Child 0133: Look, a princess. See, flowers on her dress and butterflies.
Butterfly 2: Ooh pretty. I need flowers too.
Butterfly 2 draws a flower on the dress.
Both girls get up from the table and leave the room to put their pictures in their bags.

The Anglo-Australian girls, such as Child 0133 and Princess who used Barbie and/or Princess as a reference point for their research identity, entered a gender category based on strongly normalized traditional notions of feminine beauty. This category was marked by "Barbie pink" and dresses with flowers making girls into princesses. In referencing Barbie pink, they reference princesses; and in referencing princesses, they reference whiteness. White proto-feminized ways of being produced in and through the popular culture icons they reference have become their way to understand what it means to be a "pretty" girl. Their decision to draw a girl using a pink pencil and to add the flowers and butterflies show their detailed understanding of the iconography that typically mark proto-feminized Princess and Barbie. Pink permeates all that proto-Barbie is and does. Shades of pink dominate the Disney Princess line from the graphic design of the official Disney Princess Web site to Disney's official Princesses apparel and accessory lines to its stationary, toys, and videos. Entering the official Barbie Web site you are invited to "Think pink," and the Mattel's Barbie electronics product tagline is "Plug into pink." Flowers adorn everything from Mattel's Barbie Web site to its Barbie brand product lines. Butterflies flutter through white proto-feminized Barbie's magic garden and the Barbie "Fairytopia" iconography. Child 0133 and Butterfly 2's "mundane" decision to draw a girl with flowers on her dress and butterflies with a pink pencil (re)produced their links to a proto-feminized whiteness seamlessly. That "princesses" also references whiteness becomes clear in their efforts to include the character of Pocahontas in their play.

Pocahontas is positioned within the Disney Princess product line as a Native American doll who is dressed in what Disney presents as traditional Native American clothes. Child 0133 and Princess have some familiarity with Disney Princess storylines and each is keen to find a role that enables them to play together. However, as the following excerpt highlights, they struggle with this. In their efforts to play with each other, they can't agree on Pocahontas's role in their game or on Princess's role. In an effort to find a role for Princess and Pocahontas, popular culture reference points are expanded to include Ridge and Cinderella. Princess states emphatically that Cinderella is not black and so is inappropriate for Pocahontas and the efforts to have the Pocahontas doll as a player in their game subsides. When Child 0133 proposes that Princess could be Chica—a male spider—Princess resists and asserts her desire to be a Princess who is called Child 0133.

Child 0133 is holding a Pocahontas doll.
Child 0133: I'm the little daughter and my name is Pocahontas.
Princess: I want to be the little sister.
Child 0133: No, because I . . . the little sister has to be Pocahontas.
Princess: No.
Child 0133: You be Sally (educator) and I'll be Pocahontas.
Princess: No. I don't want to be Sally.
Child 0133: Who do you want to be?
Child 0133: Ridge?
Princess: No.
Child 0133: No, because this is Australia.
Princess: What about Cinderella?
Child 0133: Cinderella's not black!
Princess: Please?
Child 0133: Do you want to be Chica?
Princess: What?
Child 0133: Chica is a spider.
Princess: I don't know what you're talking about.
Child 0133: Chica is a spider. The little boy in the jungle says "Chica."
Princess: I want to be a different name. Princess.
Child 0133: I don't know a Princess name.
Princess: I could be Child 0133.
Child 0133: Yeah!

In this exchange, the cultural categories of gender-"race" are organized and reorganized by how each of the girls reference popular culture—from "Ridge" in the soap opera *The Bold and the Beautiful* (a U.S. soap opera aired in Australia), through to Disney's Cinderella—to shape being a princess. Finally, Child 0133 and Princess become princesses through becoming themselves. In this referencing process, they locate themselves and their possibilities for playing with each other firmly within a proto-feminized whiteness. Whilst their referencing process draws on diverse popular culture icons and they struggle to find their reference points for being a princess, they will readily find a character that seamlessly connects them to whiteness. When they do, Pocahontas remains excluded and her position unresolved, yet again producing the "other" as the problematic within whiteness's seamless workings.

Fractured/Missed (Mis)connections, (Dis)connections: Vietnamese-Australian Girls, Identities, and Whiteness

Several of the Vietnamese-Australian girls connected themselves with whiteness during their project self-portraiture sessions. Self-portraiture was used to talk with children about how they understood their embodied similarities and differences. Each child was asked to look at her/his face in a mirror. They were then invited to sit at a table with researchers to talk about their physical appearance. The children were asked questions such as, "Can you tell me what color your skin is?" "What color are your eyes?" "Can you tell me what color your hair is?" "What color are your lips?" and "What color is your nose?" Children were then invited to choose from a variety of skin-toned pencils and pens to draw a self-portrait. As part of this process, the children were asked to choose the color that looked most like their skin color. Some of the children placed their hand next to the pencils to match the color. Following are excerpts from three of the individual child interviews where Kylie explored skin color with the children:

Kylie: What color is your skin?
Barbie 2: White. (Chooses the melon pencil—a pencil of light yellowy orange.)
Kylie: What color are your eyes?
Barbie 2 chooses the ebony pencil—a pencil of dark brown.
Kylie: What color are your lips?
Barbie 2: Pink.
Barbie 2 then drew the nose and drew the lips in black.

Kylie: What color is your hair?
Barbie 2: Brown.
Barbie 2 then used the melon pencil to draw her body and used the peach pencil (a pencil of light red-yellow) to draw her arms and legs.

Similarly, in the following excerpt, Barbie 3 named her skin white and also chose the melon pencil to represent this.

Kylie: What color are your eyes?
Barbie 3: Black.
Kylie: What color is your skin?
Barbie 3: White. (Chooses the melon colored pencil.)
Kylie: What color is your hair?
Barbie 3: Black.
Kylie: What color is your nose?
Barbie 3: Pink.
Kylie: What color are your lips?
Barbie 3: Pink.

Also within these interviews, Charlie's Angels 2, another of the Vietnamese-Australian girls participating in the project, connected her Vietnamese-Australian friend Barbie 1 with whiteness during the portraiture sessions.

Kylie: What color is Barbie 1's skin?
Charlie's Angels 2: White. (Chooses the almond pencil—a pencil of light browny-cream.)
Kylie: What color is Barbie 1's mouth?
Charlie's Angels 2: She has blue lipstick on.
Kylie: What color hair does the Barbie 1 have?
Charlie's Angels 2: Black.

During these self-portraiture sessions, several of the Vietnamese-Australian girls named their body/skin as "white." For Barbie 2 and Barbie 3 this connected them "doubly" to whiteness. They had chosen pseudonyms that connected them to proto-feminized whiteness and in naming their skin color as white. They reconnected their own identities to whiteness to claim whiteness as their own. As we have argued earlier in this article, Barbie works to produce and seamlessly connect identities to a particular form of proto-feminized whiteness. The connection of Barbie 2, Barbie 3, and Charlie's Angels 2 through their pseudonyms to proto-feminized whiteness alongside their desire to name and represent themselves as white during the portraiture sessions challenges us to consider what they gain and lose through the specific ways in which they choose to connect to proto-feminized whiteness. Pyke and Johnson (2003) studied the pressures that Asian-American women face as they produce their femininity within the mainstream white United States culture and within their own ethnic cultures. They found that the women in their study often rejected their ethnicity and adopted "white standards of gender" (p. 51) in ways that served as "handmaiden to hegemonic masculinity" (p. 51), and argued that a complex "matrix of oppression" (p. 52) infiltrated the meanings that Asian-American women gave to their daily lives and used to construct their identities. For all of the girls in our study, the proto-feminized popular culture icons that they accessed provided them with "white standards of gender" that persistently denied the possibility of desirable beauty and pleasurable girlhood being "ethnically other" or outside of whiteness. Within this context, to doubly choose whiteness as a Vietnamese-Australian girl might hold considerable pleasure. It offers a point of access into ways of exercising power and experiencing pleasure that comes from living within the hegemonic. It may also offer a point through which to disrupt the hegemonic by claiming it as your own.

Pedagogical Implications

The Anglo-Australian girls and the Vietnamese-Australian girls in this chapter actively located themselves within proto-feminized whiteness through their choice of pseudonyms, through their self-portraiture, and during moments of play that drew on popular culture icons. This clearly raises questions about what they each gained from doing so. To what extent did they have a developed sense of a racial hierarchy in which proto-feminized whiteness is a most desirable and privileged way to be? Did they understand that in contemporary Australia "cultural citizenship clearly belongs to 'whites'" (Cutter, 2002, p. 41)? To what extent are the Vietnamese-Australian girls resisting proto-feminized whiteness by blurring it through their claim to it? To what extent are the Anglo-Australian girls asserting proto-feminized whiteness to gain its privileges? How can we work in early childhood settings to provide spaces to blur and problematize proto-feminized whiteness in ways that open possibilities for other ways of being, as children form and re-form their racialized identities and the politics they engender?

In answering these questions, one explanation and pedagogical possibility is to position Anglo-Australian girls as knowing users of the strategies of whiteness to re-create itself as supreme and the Vietnamese-Australian girls as victims of whiteness who desire its privileges. However, there are other possibilities that have different implications for how we position "other" girls within early childhood pedagogy and practice. The Vietnamese-Australian girls could be demonstrating their capacity at a very young age to read "race" as a fluid, socially constructed category that can be rewritten. In doing so, they could also be demonstrating their potential for "resistance identities" that emerge from dominant discourses but challenge them through their emergence (Sachs, 1999). Instead of being read as "victims of whiteness" they can be read as "a new ontological hybrid, world making entities" (Latour, 1997). But, what of the Anglo-Australian girls? Can they be read as resisting proto-feminized whiteness and its privileges? Finding ways to exercise power that are pleasurable, desirable, and acceptable and that reside outside of seductive power of proto-feminized whiteness as young girls is likely to be highly challenging. Choosing to claim popular culture icons and experiencing the power and pleasure they can bring is likely to be highly seductive, especially for those young girls whose capacity to exercise power in "other" ways is extremely limited. This raises interesting and challenging pedagogical questions as we try to negotiate ways of problematizing proto-feminized whiteness and create possibilities for other multiple ways of being a girl that are powerful and desirable.

To answer our questions and explore their implications for pedagogy and education requires exploration of what a proto-feminized whiteness means to different girls in different places, spaces, and times. To do this, educators can draw on the techniques of engagement with young girls' subjectivities that we did in this chapter. They can:

1. Use popular culture icons to engage the girls in identity discussions about what is normal, beautiful, and desirable for girls to do, feel, desire, permit, and change. Analyze the girls' discourses for gender-"race" hierarchies asking questions such as:
   a. What evidence is there that discourses of proto-feminized whiteness are at play in these patterns?
   b. Who is benefiting from these discourses? Who is silenced within them?
   c. How are individual girls struggling with the desire to enact discourses of proto-feminized whiteness?
2. Engage girls in deconstructing whiteness and the relations of privilege it assumes and produces by seeking resources and

sharing ideas with them that challenge notions of beauty and normality that reinforce proto-feminized whiteness as natural, permissible, or desirable for girls. (For support with using techniques of deconstruction with young children, see Mac Naughton and Williams [2008].)
3. Talk about what all children see as doable, permissible, desirable, and changeable about being "white" girls, so what children understand about proto-feminized whiteness cannot remain unseen and therefore cannot go unchallenged. In this process, engaging with popular culture icons will be especially helpful. As we have emphasized repeatedly, it is the unseen and unchallenged that allow discourses of whiteness as normal and desirable to flourish.
4. Share with colleagues the ideas in this chapter and discuss with them ways to challenge the idea that children are innocent of "race" and that "race" is innocent of gender and vice-versa.

Final Reflections

Some recent research has explored the effects of racism awareness for children and Hughes, Bigler, and Levy (2007) offer a helpful summary of this work. They identify two areas of research that have increased understandings of how racism affects young children, especially those children who are likely to be the target of racism: 1) research on the extent to which parents teach their children about racism; and, 2) the effects on children of learning from their parents about racism.

U.S. research about if and how parents teach their children about racism shows that many Afro-Caribbean parents talk to their children about racism and how to respond to it (Hughes, 2003). Additionally, Afro-Caribbean parents are more likely than European American counterparts to teach their children about racism. Whilst some white parents in the United States believe it is important to talk to their children about racism (Iyer, Leach, and Crosby, 2003), European American parents often ignore or avoid teaching their children about racism (Katz, 2003). It appears that discourses of childhood colorblindness and racial innocence are in part responsible for this view amongst parents (Hughes, Bigler, and Levy, 2007; Norton et al., 2006). It may be for this reason, that there is scant research about the effects of learning about racism on white children (Meltzer, 1939). From the research with U.S. Afro-Caribbean parents its clear that young children can engage productively in discussions about racism from an early age. We believe that young girls are not born understanding racial hierarchy and racial discrimination, or with some inherent skills in knowing how to act within hierarchies of beauty. Instead, they learn this through referencing themselves to others as they engage with the world in the process of forming and re-forming their identities. Early childhood spaces are one place where their negotiations of gendered identities are constructed and reconstructed and the politics of effects of this lived. Given the international commercial success of the Barbie, Disney Princesses, and Fairies product lines, they clearly embody proto-feminized whiteness in ways that are highly seductive and desirable for many young girls beyond the borders of Australia. For instance, various Web-based sources estimate that Barbie is sold in over 150 countries grossing over US$150,000 billion per annum and that the Disney Princesses are sold in over 90 countries with a sales turnover of around US$4.5 billion annually.

Against the diversity of their sales base, the storylines of these global cultural products for children provide increasingly restricted entry points for girls from diverse ethnicities and nationalities to embody ways of being beautiful, and instead produce points of tension for girls who seek to claim and use and/or subvert them in their daily lives. Whilst young girls may or may not be aware of the specific privileges that accrue to them by referencing themselves within a proto-feminized whiteness, for the girls in our study, such culturally constructed and mediated reference points were embedded in their racialization and were inseparable from it. How we work to challenge and reinscribe these negotiations and constructions for young children, especially girls, in our early childhood spaces becomes important in order to problematize and deconstruct dominant white femininities as linked to beauty, and instead work to create ways of being a girl that embrace multiple images of being a girl.

Taking up these challenges is important if we are to effectively target antiracist education to local children's subjectivities against global cultural icons and to engage them proactively in remaking what children do, permit, desire, and change in building their gender-"race" identities with us in early childhood education. Challenging the negative effects of the racism that proto-feminized whiteness brings for children of color is critical in this process. Hughes, Bigler, and Levy (2007) summarize several studies (e.g., Hughes, 2003) that focus on the effects for U.S. Afro-Caribbean children learning about racism from their parents. The positive effects they noted were that these children have better grades in school, fewer behavioral problems, and strongly developed racial identities. However, negative effects for children have also been reported (Caughy, Nettles, O'Campo, and Lohrfink, 2006). A study of the different outcomes of racial awareness teaching by parents, Caughy et al. (2006) found higher levels of behavioral problems for girls and higher level of externalizing problems for boys. Additional negative effects reported were depression and stress. These effects result from children knowing that racism will be directed toward them. However, those studies have been primarily focused on youth rather than young children (Hughes, Bigler, and Levy, 2007). Whilst, this suggests some potential risks in talking about the negative effects of proto-feminized whiteness with young children of color, we believe that silence brings its own risks and harm that need acknowledging and addressing.

Further Resources

For those early childhood educators and readers of this book who would like to further explore antiracist pedagogies in their own practices, the resources listed below provide you with some starting points. There are a number of texts that can help you explore pedagogical approaches as well as a number of texts you could use with young children.

Mac Naughton, G., and G. Williams. (2008). *Teaching techniques for young children*, 3rd ed. Melbourne: Pearson Education.

Robinson, K.H., and C. Jones Diaz. (2006). *Diversity and difference in early childhood education: Issues for theory and practice.* England and New York: Open University Press.

Children's Books

Coleman, E. (1996). *White socks only.* Illinois: Albert Whitman and Company.

Hoffman, M. (1991). *Amazing Grace.* New York: Dial Books for Young Readers.

——— (1995). *Boundless Grace.* New York: Puffin Books.

Medearis, M., and A.S. Medearis. (2000). *Daisy and the doll.* Vermont: The Vermont Folklife Center.

CHAPTER 5

Masculinities, Mateship, and Young Boys: Defending Borders, Playing Footy, and Kissing Meg

Karina Davis and Glenda Mac Naughton

Introduction

In chapter 4 we explored how whiteness constructed exclusions and possibilities in the "racing" of young girls' identities. In this chapter, we turn attention to how white masculinities construct the "racing" of young boys' identities. In particular, we look at how discourses of white, hegemonic masculinity create dynamics of inclusion and exclusion in an early childhood setting and how they can be used to sanction white boys who step outside of those discourses. Given discourses shape our sense of who we are, discourses of whiteness and masculinity embody a politics of identity that shape what boys understand what they can do, think, and feel, what they are permitted to do, think, and feel, what they should desire, and what they can change.

To map the specific political dynamics of discourses of whiteness and masculinity in shaping the gendered-"race" identities of young boys, we again use critical race theories (Delgado and Stefancic, 1997), specifically critical white theories (Frankenberg, 1993), and postmodern perspectives on identity as chosen and performed rather than given and fixed (Hage, 1998; Mansfield, 2000; Pease, 2004). We highlight how racism is a part of both children's everyday and institutionalized social life in an early childhood space and how male gendered whiteness is a cultural destination that is often invisible and normalized (Frankenberg, 1993; Moreton-Robinson, 2004) in the daily lives of children. We reiterate the argument we built in chapter 4 that the politics of children constructing their identities through discourses of whiteness is neither concrete nor fixed. Instead, boys' enactment of and embodiment in discourses of whiteness shifts and changes as contexts change and their desires shift. We show how targeted observations of the dynamics of children's relationships in play, analyzed using critical race theories, can support educators to disrupt the complex dynamics of racism in the lives of young children in early childhood settings by making "race"-power seeable and seen in young children's identity construction.

Masculinities, Hegemony, and Australia

Boys build a sense of who they are, who they should be, and who they want to be as boys through gendered discourses of masculinity. In the field of gender studies, there is considerable debate about how the construction of masculinity and masculinities work (Connell, 1998; Demetriou, 2001; Newton, 1998). In the 1990s, supported by the wider acceptance and dissemination of poststructural, postmodern, and feminist theories of identity construction (Newton, 1998), theorizing masculinity incorporated acknowledgment and discussion of multiple discourses of masculinity. The term "masculinities" arose from this work to emphasize the many ways of being masculine and to challenge the idea that there is one normal, fixed, and mature masculine identity that all boys should aspire to. This represented a shift in thinking from the idea that there is one way to be male to the idea of men and boys constructing a sense of self through multiple discourses of masculinities. The choices men and boys made was understood to be influenced by their historical, social, and geopolitical location and their "race," class, and sexuality identities. This made different and multiple masculinities possible as Wang (2000) discussed:

> According to feminist and other scholars, masculinity is historically and socially constructed...[T]he form practiced by the dominant group in a culture is called "hegemonic masculinity"...[B]ecause other forms of masculinity can coexist with it, a boy can draw on multiple masculinities in his efforts to develop his identity both as an individual and as a male. (p. 114)

In what follows, we show how masculinities intersect and are inextricably bound with "race." Boys are shown taking up and using inclusions and exclusions constructed through and within white hegemonic masculinities within their group in one of the centers in the PCCRCD project (refer appendix). They negotiated and struggled with enacting the ideals of hegemonic masculinities and we point to the implications for young boys in these negotiations and struggles as they created, maintained, and/or were denied a place of power in the hegemonic hierarchy of relationships in their group (Renold, 2004; Wang, 2000). We understand hegemonic masculinity as follows:

> The type of masculinity the dominant group performs is called hegemonic masculinity. Hegemonic masculinity is a "culturally idealized form of masculine character." (Connell, 1990, p. 83)
>
> Hegemonic masculinity then is the current ideal. (Cheng, 1999, p. 3)

In drawing from this definition, we acknowledge that the "idealized masculine character" is contingent. It shifts and changes over time and across social and cultural groups and contexts. In the Australian context, as in many other patriarchal nation states, hegemonic masculinity is constructed in a particular and geohistorical specific image of whiteness and against identities of many "others"—including feminine, nonheterosexual, racial, and cultural others. More specifically, hegemonic Australian white masculinities are tied inherently to discourses of "mateship." Mateship discourses permit and desire for men and boys "physical prowess, loyalty to mates and additionally for men the fulfilment their roles as 'breadwinner' in direct opposition to the role of women as nurturers" (Murrie, 1998).

In Australia, the intersections of hegemonic masculinities with concepts of mateship construct a historically and geopolitical masculine ideal in the image of the white male that has been "elevated to national traditions" (Murrie, 1998). This concept of mateship was forged through white settler mythology and later solidified through stories of war and exploits of Australian soldiers (Garton, 1998). Mateship is understood as the celebration of male friendship. It embraces homo-social bonds inclusive only of white males and it valorizes antiauthoritarianism and loyalty to mates (Haltof, 1993; Hatchell, 2003; McGuffey and Rich, 1999; Murrie, 1998). In this white Australian ideal of hegemonic masculinities, those boys who enact the image of the settler and war hero—strong, brave, and adventurous—with the ideals of mateship are accorded a position of power and privilege. These Australian discourses of mateship are also seen to be coercive (Cowlishaw, 2004) and contingent on male peer approval of displays to validate masculinity (Murrie, 1998). In these ways, displays of hegemonic masculinities must be continually and publicly enacted and performed in order to validate them. Using a case study from the PCCRCD, we show the existence and location of these displays and their "racialized" exclusionary effects in an early childhood space.

Observation as a Tool for Uncovering White Hegemonic Masculinities

Observation in the early childhood field has long been seen as an important, if not essential, tool for the adult to document and chart a child's understandings, growth, and learning (Nilsen, 1997; Rolfe, 2001). Individual and group observations are highly valued for the information they uncover about children. For this reason, educator's, educator-researchers, and researchers use these tools (Dau, 1999; Rolfe, 2001). However, the dominance of developmental psychological understandings of the child in giving meaning to these observations can silence the existence and enactment of racism and discrimination at individual, group, and institutional levels. It also overlooks the intersections of racism between the individual, group, and institutions (Cannella, 1997; Lubeck, 1994). Within the PCCRCD project we aimed to disrupt silencing "race" discrimination through using research methods and tools that were designed to uncover "race" and its significance in children's lives.

In this chapter we draw on the field-based observations we made in the PCCRCD study. In these observations we targeted mapping group dynamics, power relationships, and friendship patterns between children over an eighteen-month period. We designed an observation tool and protocol that enabled us to track the incidences and patterns of children being positioned as leaders, participants, and/or excluded from groups during free play periods and to chart the social dynamics of the group (adults and children) as a whole. In addition to observer-generated data, the researchers checked the meaning and significance of specific incidents and patterns with the children and adults. We analyzed these patterns using critical race theory and postmodern theories of discourse analysis to focus our attention on how discourses of whiteness and hegemonic masculinity worked in tandem to produce moments of racism. In what follows we illustrate this via a case study of Zurg (a child-chosen research pseudonym) to explore how dominant whiteness intersected and was bound up within the enactment and enforcing of hegemonic masculinities and how this established gender-"race" hierarchies in the group.

Masculinities Intersecting with Dominant White Identities: Snapshots of Zurg

Zurg was three years old when the PCCRCD study began. He was positioned by the staff within his children's service and by Karina as researcher early in the project, as a leader within the group. He was articulate, social, and humorous, at times displayed sensitivity and gentleness and appeared to have well-developed friendships both within his core group of white boys, as well as across other social groups within the service. These very early understandings were based on general early childhood observational processes designed to assess how individual children were developing based on psychological understandings of the child. These observational processes however did not uncover or "see" how Zurg worked to exclude and silence others or how he asserted his authority within the group through sanctioning and asserting particular discourses of hegemonic white masculinities. Further, how Zurg's peers used these understandings of white masculine discourses to enforce conformity for Zurg at times also remained unseen within this practicing of early childhood curriculum and assessment. It was employing the theoretical grounded approach of reading the data from within understandings of whiteness and hegemonic masculinities that made it possible to recast Zurg's observations in order to raise new and different questions about his actions and relationships within his early childhood space.

In the following three vignettes (taken over the eighteen months of data collection for the PCCRCD project), we explore how Zurg and his peers experience and produce themselves through discourses of white hegemonic masculinities in an inner-Melbourne, Australian early childhood space.

Zurg as Boundary-Defender

Vignette 1

Zurg is playing outdoors in a house constructed by him, James, A Cow, and Andy. Ricky is on the fringes of the house watching these four boys. Zurg says to others as they rush in and out of the door of the house blocking access to others "Let's guard our house and don't let them in."

In Vignette 1, Zurg was playing in a corner of the outside area with a small group of three- and four-year-old white boys. Zurg and these white boys appeared to have an established group and actively sought each other out across their time at the center. This group had gathered blocks from the outside areas and built a "house" in the corner of the playground. Ricky, another white male who was never an active member of their play or group but often hovered on the fringes and often appeared to be ignored by the boys in the group, was sitting on a block close by. The other children from the group were also playing outside and were however engaged in a variety of other activities.

Each of the boys at this center was firmly positioned within the politics of his community, which arguably reflects the politics of a nation (Hage, 1998; Wadham, 2004). In Vignette 1 Zurg and his peers are positioned as powerful and dominant through how they embodied particular ways of being both white and male—one of the implications of this was the power to claim, guard, and defend their territory. While Frankenberg (1997) discusses that "like white women, white men are diversely located in relation to power and privilege" (p. 13), Zurg's powerful location is illuminated through his active enforcing and reinforcing of spatial boundaries and claims of ownership, both for himself and alongside his white male peers. While this active enforcement has strong parallels with Taylor's (2005) study in which she found young white boys actively claimed land, it also sits alongside Australia's history and present, which is littered with accounts of the white male conquering adversity and defending borders while building an image of White Australia as a "masculinist ideal...made in the image of ruling white men who dominate the processes and practices of nation building" (Wadham, 2004, p. 192).

Critical white theorists argue that the white male idealization provides particular sorts of possibilities for the "right" white male to not only assert, but to expect, rights of ownership (Frankenberg, 1997; Taylor, 2005). Layering these theoretical understandings into how we make sense of the house blockade inserts politics into understanding this "play" by pointing to its intersections with larger "race" and diversity. This event was not unusual. Zurg and his white male peer group often claimed land and blockaded it. They were the only group within the setting to actively build and enforce boundaries. Did their enactment and performance of powerful hegemonic masculinities make their claims for land possible? If so, in what ways was this made possible?

Discourses of hegemonic masculinities construct hierarchies for boys in which the boys who enact the hegemonic (and idealized) masculinity are placed at the top of the hierarchy. From this position, the boys can manipulate the hierarchy to ensure they remain in powerful positions (McGuffey and Rich, 1999). For Zurg and his "mates," the enactment of specific hegemonic white masculinities valorized in their context accorded them power and privilege to claim space, a valuable commodity, in their communal setting, and continually reclaim their place on the top of hierarchy through this claim to land. The existence of a hierarchy of masculinities is further illustrated in the boys' active ignoring of Ricky—a young white boy who enacted a different nonhegemonic masculinity. Ricky was introverted, slight, and spoke quietly, but often positioned himself just within or very close to the territory of Zurg and his mates. However, the border defenders did not chase Ricky out of their borders. They did chase the girls out and they did chase another boy who was "racially other" to them if he or the girls accidentally or actively tried to breach their borders. This hierarchy of responses supports and rearticulates white hegemonic masculinities as powerful and makes possible the grudging tolerance of the "wrong" sort of white male and active exclusion of gendered and racial others.

This power to exclude others, claim ownership, and reinforce specific masculinities is further highlighted in the second vignette when once again the group were playing outside and Zurg had actively sought out Meg—the white educator in the group.

Zurg as Performing Heterosexism

Vignette 2

> Zurg has been observed for the fourth time now outside sitting on Meg's lap and talking intently with her. They laugh and exchange affectionate hugs, hold hands and Zurg uses his hand gently to turn Meg's face to his when her attention wanders away from him. For me, this observation sits over the top of Zurg's interview data where he talks about Meg as being a highly desirable friend and of the appropriateness of kissing her as opposed to kissing his other male friends. It also intersects with my overheard conversations where Zurg claimed Meg as his girlfriend—a claim undisputed by his peers in this discussion.

Hegemonic masculinity in Australia rests heavily on "projecting an overt compulsory heterosexuality" (Renold, 2004, p. 249). As has been mentioned earlier, hegemonic masculinities also require public displays of masculinities in order to enforce and reinforce one's position in the masculinities hierarchy (McGuffey and Rich, 1999). With these understandings in mind, Vignette 2 can be read as Zurg publicly claiming female Meg in order to confirm his white hegemonic masculinity to his peers. This allows him to assert and reinforce his dominant place in the hierarchy to and within the group.

Within Australia, discourses of masculinities tie hegemonic white masculinities with heteronormative and homosocial ideals and in turn they tie Australian white males together in complex relationships of "mateship" (Cowlishaw, 2004; Hage, 1998; Murrie, 1998; Pease, 2004). Martin (2004) argues that "women feature in male homosociality as items of exchange, and proof of a masculinity actually validated by other men in the homosocial bonds" (p. 37). Zurg's ability to "claim" Meg and have this recognized by his white peers as a relationship with credibility (as discussed in his claiming if Meg as his girlfriend) and authenticity rests on his status as a dominant white male and, conversely, it is this relationship with Meg that reinforces and validates his position. Using theories of whiteness and masculinities in strategic and political ways to read the observations we took during the project also made it possible for us to see Zurg's acts of kissing and claiming Meg as enactment of hegemonic masculinities and strategic acts of reasserting his dominant heteronormative hegemonic white masculinities. These acts however, were given authenticity and credibility and worked to confirm his dominant position because of his already confirmed and rearticulated dominant status within the group.

While Zurg could often be seen constructing and being constructed by hegemonic masculinities in powerful ways, it was also possible to see how this positioning was fluid and required ongoing negotiations by Zurg to maintain this position. In Vignette 3 this active and ongoing negotiation and the sanctioning for Zurg as he steps outside of hegemonic masculinity practices is highlighted.

Challenging Dominant White Masculinities: Zurg and the Dolls

Vignette 3

Zurg has been really interested in the Persona Dolls. On the first day I brought them in he was keen to help me set them up and asked many questions about them. In his first interview he stayed with me to help me pack them away and within this he showed great care and gentleness in how he handled the Dolls. The times I have brought them outside so the children could interact with them,

> he continually returned to me and the Dolls and sanctioned others if they were treating them roughly. All of this has been interesting to watch especially the reactions and strategies used by the other white boys to draw him back into their games and away from the Dolls. Today he was playing with the Persona Doll's and calls to James to join him. James says "No." James, A Cow and another child consult and have a whispered conversation while watching Zurg with the dolls. They search the outside area until they find a football and call out to Zurg to join them. Zurg puts Aravinda down and runs to join the others and a game of football is fiercely contested over the entire outside space for the next 15 minutes. It is interesting to note that only the white boys are invited to and participate in this game.

Discourses of hegemonic masculinities and the practices associated with them are more often defined by what these masculinities reject, denigrate, or are disassociated from (Epstein, 1997; Martino, 1999; Renold, 2004). In these discourses, hegemonic masculinities work to disassociate from feminine and nurturing behaviors and it is argued by Epstein (1999) that within these constructions of masculinity "to feminize a man is to disempower him" (p. 13).

When Zurg steps outside of practicing hegemonic white masculine behaviors and into behaviors more akin to nurturer (playing with the Persona Dolls), his white male peers employ overtly masculine cues (the playing of Australian Rules football) to draw Zurg back into dominant discourses of being a white hegemonic male. Within understandings of Australian specific hegemonic masculinities, it is especially powerful that Zurg's peers chose to draw him into more public displays of dominant hegemony with a sports game specific to and important to the local dominant culture of Melbourne. Skill at, and knowledge of, sport is important to the formation of hegemonic masculinities (Burgess, Edwards, Skinner, 2003) and within an Australian context, football is seen to reinforce and represent these masculinities (Martino, 1999). That the game of football succeeded in drawing Zurg out of his display of alternative masculinity and the game itself then proceeded to claim a large amount of the outside space illustrates the success of this tactic in both sanctioning and drawing Zurg back into dominant displays of white hegemonic masculinities. Sitting next to this game is the complete exclusion of others in it, even though one nonwhite boy is highly skilled at football and has extensive knowledge of the rules.

Exploring Vignette 3 using discourses of white hegemonic masculinities, we see that Zurg's identity as a dominant white male is not fixed and concrete and he is constrained and sanctioned by the dominant white hegemonic masculinities that he often embodies. Zurg's displays of alternative discourses of masculinity did not fit within his white male peer group's discourses of acceptable white hegemonic behavior. His peer group was quick to sanction his behavior, drawing on well-established and highly gendered Australian masculine cues. In this moment, Zurg faced peer disapproval for his embodied masculinities in ways that worked to assert and reinforce dominant white hegemonic masculinities over him. Further, white hegemonic and exclusive masculinities were performed in the spectacle of the game through the active exclusion of others.

Reading for Dominant Masculinities and "Race"

Reading these three vignettes as specific and isolated observations would not have privileged the influence and salience of "race" and gender within Zurg's understandings and practices in his children's service. Nor would they have pointed to the racism apparent in how dominant masculinities were produced. As Copenhaver-Johnson (2006) writes of the silenced and hidden influence of "race" in the constructions of girls' friendships, "they (children) did not explicitly speak of race. No single example from my notes, in isolation, would be convincing of a pattern of behaviour" (p. 16). However, similar to Copenhaver-Johnson's argument, when Zurg's observations were layered over each other and then further layered with observations and events from Zurg's peers and other children within his group, patterns of behavior with racist effects within broader group dynamics were tracked. Analyzing these patterns using critical race theories of whiteness and gender theories of hegemonic masculinities enabled increasingly complex racialized and gendered discourses and their effects to be seen and named. In Zurg's case, this reiterative layering illuminated the enforcing and reinforcing of dominant hegemonic white masculinities within the group. Within this, Zurg was located as multiple and contradictory—simultaneously enforcing specific and desirable white hegemonic masculinities that excluded many others while being constrained and controlled by these himself. Whilst Zurg is agentic in choosing identities that work for him, these choices are always made in a discursive field where the competition between discourses makes some of his choices more permissible, desirable, and doable than others.

Pedagogical Implications

Epstein (1997) argued that schools are "sites of masculinity making" (p. 4). If we are to acknowledge and accept that in early childhood spaces, "racialized" hegemonic masculinities are being struggled over, negotiated, and used to include and exclude, then finding ways to enter into these negotiations with young boys is important antiracist work. In antisexist work, Renold (2004) asks that educators pay attention to the dynamics of masculinity, particularly how alternative masculinities are constructed and enacted, in order to find points of challenge to hegemonic masculinities. In antiracist/antisexist work it is important to explore how hegemonic masculinities are "raced" in order to expand the antiracist possibilities that young white boys explore for themselves and permit others. There are some clear strategies educators can use in this work to develop pedagogical responses to the politics of children's "race"-gender identities.

1. Observe the power dynamics of the group of children over time. Use these observations to identify who is able to take the role of leader and who is excluded from this role. Analyze the patterns for gender-"race" hierarchies asking questions such as:
    a. What evidence is there that discourses of white hegemonic masculinities are at play in these patterns?
    b. Who is benefiting from these discourses? Who is silenced within them?
    c. How are individual boys struggling with the desire to enact discourses of white hegemonic masculinity?
2. Engage children in deconstructing whiteness and the relations of privilege it assumes and produces by challenging that what happens between white boys and boys of color and white boys and all girls is natural, permissible, or desirable. Talk about what children see as doable, permissible, desirable, and changeable about being "white" boys, so what children understand cannot remain unseen and therefore go unchallenged. It is the unseen and unchallenged that allow discourses of whiteness as normal and desirable to flourish. (For support with using techniques of deconstruction with young children see Mac Naughton and Williams [2008].)
3. Talk with colleagues about the ideas in this chapter to challenge the idea that children are innocent of "race" and that "race" is innocent of gender and vice-versa.

Final Reflections

In discussing efforts to challenge the unjust effects of hegemonic masculinities in schools, Renold (2004) argues that:

> If we are seeking to transform the oppositional and hierarchical gender constructions that permeate children's identity, work and peer relations, then schools and school policies need to pay equal attention to the margins and the center. Exploring (especially with children themselves) how they are interconnected (and indeed interact with social class, ethnicity, religion, sexuality and age) may well go some way to disrupting the power relations that constitute the gendered hegemonic matrix that all children (boys and girls) negotiate on a daily basis. (p. 262)

Racism relies on complex relations of power being maintained at the site of the individual and the institution. Zurg's case study reminds us that the complexity of racism is linked to the complexity and politics of children's identity choices and these choices in turn are linked to discourses we make available to them and challenge.

Early research (e.g., P. Katz and Zalk, 1978; Katz, 1973) about children unlearning racism studied very specific and precise strategies in test conditions. For instance, in one study children were taught specific names for people of other "races" using a series of photographs. They measured the effects of this cognitive training on their levels of prejudice and compared those levels with what happened when children observed the faces without being given labels and when asked to make same-different judgments about the faces (Katz, 1973). Children who observed after were given specific names for the people in the photographs displayed the lowest level of prejudice. More recent U.S. research suggests that critically engaging young elementary school students with discourses about historical racism did change how they understood "race" (Hughes, Bigler, and Levy, 2007). In Northern Ireland there is also strong and promising evidence from studies with over 3,500 children in over 200 early years settings that programs to reduce young children's intergroup and ethnic prejudice can work (e.g., Connolly, Fitzpatrick, Gallagher, and Harris, 2006). In this work is strong support and hope for efforts by educators to deconstruct whiteness as part of antiracist education in the early years.

Further Resources

The resources listed below may be beneficial to some readers and educators, especially those unfamiliar with antiracist and antisexist pedagogies, in considering how they might place gender-"race" into account in their pedagogies. There are a number of texts focused on pedagogies and curriculum listed. However, there are not as many texts for children provided below, as children's texts locating young boys in "race" negotiations were (and are) difficult to source.

Further Resources

Dau, E. Ed. (2001). *The anti-bias approach in early childhood*, 2nd ed. Frenchs Forest, NSW: Longman.

Mac Naughton, G., and G. Williams. (2008). *Teaching techniques for young children*, 3rd ed. Melbourne: Pearson Education.

Children's Books

hooks, b. (2002). *Be boy buzz.* New York: Hyperion Books for Children.

———. (2004). *Skin again.* New York: Hyperion Books for Children.

Walter, M.P. (2004). *Alec's primer.* Vermont: Vermont Folklife Center.

Part I References

Aboud, F.E. (1988). *Children and prejudice.* UK: Basil Blackwell.

Aboud, F.E., and A. Doyle. (1996). Does talk of "race" foster prejudice or tolerance in children? *Canadian Journal of Behavioural Science*, 28(3), 161–170.

Aboud, F.E., and M. Amato. (2001). Developmental and socialization influences on intergroup bias. In S.L. Gaertner (Ed). *Blackwell handbook of social psychology: Intergroup processes.* Oxford: Blackwell.

Alvermann, D. (2000). Researching libraries, literacies and lives: A rhizoanalysis. In E. St. Pierre and W. Pillow (Eds). *Working the ruins: Feminist poststructuralist theory and methods in education* (pp. 114–129). Routledge: London.

Ang, I. (2003). Together in difference: Beyond diaspora, into hybridity. *Asian Studies Review*, 27(2), 141–154.

Apfelbaum, E.P., K. Pauker, N. Ambady, S.R. Sommers, and M.I. Norton. (2008). Learning (not) to talk about race: When older children underperform in social categorization. *Developmental Psychology*, 44(5), 1513–1518.

Ashcroft, B. (1994). *The Post-colonial studies reader.* London: Routledge.

Atkinson, J. (2002). *Trauma trails recreating song lines.* Melbourne: Spinifex Press.

Averhart, C.J.B., and R.S. Bigler. (1997). Shades of meaning: skin tone, racial attitudes, and constructive memory in African American children. *Journal of Experimental Child Psychology*, 67, 363–388.

Baker-Sperry, L., and L. Grauerholz. (2003). The pervasiveness and persistence of the feminine beauty ideal in children's fairy stories. *Gender and Society*, 17(5), 711–726.

Bennett, D.E. (1998). *Multicultural states: Rethinking difference and identity.* London: Routledge.

Best, D., C. Naylor, and J. Williams. (1975). Extension of color bias research to young French and Italian children. *Journal of Cross-Cultural Psychology*, 6, 390–405.

Best, D., J. Field, and J. Williams. (1976). Color bias in a sample of young German children. *Psychological Reports*, 38(3), 1145.

Bigler, R.S., and L.S. Liben. (1993). A cognitive-developmental approach to racial stereotyping and reconstructive memory in Euro-American children. *Child Development*, 64(5), 1507–1518.

Black-Gutman, D., and F. Hickson. (1996). The relationship between racial attitudes and social-cognitive development in children: An Australian study. *Developmental Psychology*, 32(3), 448–456.

Bogardus, E. (1925). Social distance and its origins. *Journal of Applied Sociology*, 9, 299–308.

Bourdieu, P. (1990). *The logic of practice.* Cambridge: Polity Press.

Brand, E.S., R.A. Ruiz, and A.M. Padilla. (1974). Ethnic identification and preference: A review. *Psychological Bulletin*, 81(11), 860–890.

Bray, M. (2002). Skin deep: Dying to be white. Sourced at http://edition.cnn.com/2002/WORLD/asiapcf/east/05/13/asia.whitening/ on May 15.

Brinson, S. (2001). Racial attitudes and racial preferences of African-American preschool children as related to the ethnic identity of primary caregivers. *Contemporary Education*, 72, 15–21.

Burgess, I., A. Edwards, and J. Skinner. (2003). Football culture in an Australian school setting: The construction of masculine identity. *Sport, Education and Society*, 8(2), 199–212.

Buswell, G.T. (1929). Review: [untitled]. *The Elementary School Journal*, 30(3), 233–234.

Butler, J. (1997). *Excitable speech: A politics of the performative.* New York: Routledge.

Cameron, J.A., J.M. Alvarez, D.N. Ruble, and A.J. Fuligni. (2001). Children's lay theories about ingroups and outgroups: Reconceptualizing research on prejudice. *Personality and Social Psychology Review*, 5(2), 118–128.

Cannella, G.S. (1997). *Deconstructing early childhood education. Social justice and revolution.* New York: Peter Lang.

Caughy, M.O., S.M. Nettles, P.J. O'Campo, and K. Lohrfink. (2006). Neighbourhood matters: Racial socialization and the development of young African American children. *Child Development (Special Issue on Race, Ethnicity and Culture)*, 77(5), 1220–1236.

Cheng, C. (1999). Marginalised masculinities and hegemonic masculinity: an introduction. *Journal of Men's Studies*, 7(3), 295–310.

Chin, E. (2001) *Purchasing power: Black kids and American consumer culture.* Minnesota: University of Minnesota Press.

Choules, K. (2006). Globally privileged citizenship. *Race, Ethnicity and Education*, 9(3), 275–293.

Clark, K., and M. Clark. (1939). The development of consciousness of self and the emergence of racial identification in Negro preschool children. *Journal of Social Psychology, S.P.S.S.I. Bulletin*, 10, 591–599.

Coloma, R.S. (2006). Disorienting race and education: Changing paradigms on the schooling of Asian Americans and Pacific Islanders. *Race, Ethnicity and Education*, 9(1), 1–15.

Connell, R.W. (1990). The state, gender and sexual politics: theory and appraisal. *Theory and Society*, 19, 507–544.

———. (1998). R.W. Connell's "masculinities": A reply. *Gender and Society*, 12(4), 474–477.

Connolly, P. (2000). Racism and young girls' peer-group relations: The experiences of South Asian girls. *Sociology*, 34(3), 499–519.

———. (2006). The masculine habitus as "distributed cognition": A case study of 5- to 6-year-old boys in an English inner-city, multi-ethnic primary school. *Children and Society*, 20, 140–152.

Connolly, P., S. Fitzpatrick, T. Gallagher, and P. Harris. (2006). Addressing diversity and inclusion in the early years in conflict-affected societies: A case study of the media initiative for children—Northern Ireland. *International Journal of Early Years Education*, 14(3), 263–278.

Copenhaver-Johnson, J. (2006). Talking to children about race. The importance of inviting difficult conversations. *Childhood Education*, 83(1), 12–22.

Cowlishaw, G. (2004). *Blackfellas, whitefellas and the hidden injuries of race.* Carlton, Victoria: Blackwell.

Criswell, J.H. (1937). Racial cleavage in Negro-white groups. *Sociometry*, 1, 81–89.

———. (1939). A sociometric study of race cleavage in the classroom. *Archives of Psychology*, 33, 1–82.

Cutter, M. (2002). Empire and the mind of the child: Sui Sin Far's "Tales of Chinese children"—critical essay. *MELUS*, 27(2), 31–48.

Dau, E. (1999). "I can be playful too": The adult's role in children's socio-dramatic play. In E. Dau (Ed). *Child's play. Revisiting play in early childhood settings* (pp. 187–201). Sydney, NSW: MacLennan and Petty.

Davies, B. (2001). Eclipsing the constitutive power of discourse: The writing of Janette Turner Hospital. In E.A. St Pierre and W.S. Pillow (Eds). *Working the ruins: Feminist poststructuralist theory and methods in education* (pp. 179–198). New York: Routledge.

Davis, S., P. Leman, and M. Barrett. (2007). Children's implicit and explicit ethnic group attitudes, ethnic group identification, and self-esteem *International Journal of Behavioral Development*, 31(5), 514–525.

Deleuze, G., and F. Guattari. (1987). *A thousand plateaus: capitalism & schizophrenia.* London: Athlone Press.

Delgado, R., and J. Stefancic. Eds. (1997). *Critical white studies. Looking behind the mirror.* Philadelphia: Temple University Press.

Demetriou, D.Z. (2001). Connell's concept of hegemonic masculinity: A critique. *Theory and Society*, 30(3), 337–361.

Dikotter, F. (1994). Racial identities in China: Context and meaning. *The China Quarterly*, 138 , 404–412. Sourced at http://links.jstor.org/sici?sici=0305-7410%28199406%290%3A138%3C404%3ARIICCA%3E2.0.CO%3B2A on August 21, 2008.

Dohnt, H., and M. Tiggemann. (2006). Body image concerns in young girls: The role of peers and media prior to adolescence. *Journal of Youth and Adolescence*, 35, 141–151.

du Cille, A. (1996). Dyes and dolls: Multicultural Barbie and the merchandizing of difference. In J. Munns and G. Rajan Longman (Eds). *A cultural studies reader: history, theory, practice.* New York: Longman.

Epstein, D. (1997). Taking it like a man: Narratives of dominant white masculinities in South Africa. Paper presented at the Australian Association for Research Conference, Brisbane, November 30–December 4.

Feinman, S. (1979). Trends in racial self-image of black children: Psychological consequences of a social movement. *The Journal of Negro Education*, 48(4), 488–499.

Fleming, P. (2002). "Lines of flight": A history of resistance and the thematic of ethics, death and animality. *Ephemera: Critical Dialogues on Organization*, 2(3), 193–208.

Foucault, M. (1980). *Power/knowledge: Selected interviews and other writings.* New York: Pantheon Books.

———. (1985). *The history of sexuality volume 2.* New York: Pantheon Press.

Frankenberg, R. (1993). *White women, race matters. The social construction of whiteness.* Minneapolis: University of Minnesota Press.

———. (1997). Introduction. Local whitenesses, localising whiteness. In R. Frankenberg (Ed). *Displacing whiteness. Essays in social and cultural criticism* (pp. 1–34) Durham: Duke University Press.

Frederick, R. (1927). An investigation into some social attitudes of high school students. *School and Society*, 22, 235–242.

Garton, S. (1998). War and masculinity in twentieth century Australia. *Journal of Australian Studies*, 56, 86–96.

Ghandi, L. (1998). *Postcolonial theory: A critical introduction.* Sydney: Allen and Unwin.

Goodman, M. (1964). *Race awareness in young children.* Massachusetts: Collier.

Gopaul McNicol, S. (1995). A cross-cultural examination of racial identity and racial preference of preschool children in the West Indies. *Journal of Cross-Cultural Psychology*, 26(2), 141–152.

Grosz, E. (1990). Contemporary theories of power and subjectivity. In S. Gunew (Ed). *Feminist knowledge: Critique and construct* (pp. 59–120). London: Routledge.

———. (1994). *Volatile bodies: Toward a corporeal feminism.* Sydney: Allen and Unwin.

Hage, G. (1998). *White nation: Fantasies of white supremacy in a multicultural society.* Annandale: Pluto Press.

Hall, K. (1999). White feminists doing critical race theory: Some ethical and political considerations. *APA Newsletters*, 98(2).

Haltof, M. (1993). . . . in quest of self-identity . . . *Gallipoli*, mateship and the construction of Australian national identity. *Journal of Popular Film and Television*, 21(1), 27–36.

Hamman, R.B. (1996). Rhizome @ Internet: Using the Internet as an example of Deleuze and Guattari's "rhizome." Sourced at http://www.socio.demon.co.uk/rhizome.html on August 24, 2008.

Hanks, W.F. (1996) *Language and communicative practices.* Boulder: Westview Press.

Harrison, A., and S. O'Neill. (2003). Preferences and children's use of gender-stereotyped knowledge about musical instruments: Making judgements about other children's preferences. *Sex Roles: A Journal of Research*, 49, 389–400.

Hatchell, H. (2003). Masculinities and whiteness: The shaping of adolescent male students' subjectivities in an Australian boys' school. Unpublished PhD thesis: Murdoch University.

Hewstone, M. (1988). Attributional bases of intergroup conflict. In W. Stroebe, A.W. Kruglanski, D. Bar-Tal, and M. Hewstone (Eds). *The social psychology of intergroup conflict: Theory, research and applications* (pp. 44–71). New York: Springer.

Hirschfeld, L. (1995). Do children have a theory of "race"? *Cognition*, 54, 209–252.

Hitlin, S., J.S. Brown, and G.H. Elder. (2006). Racial self-categorization in adolescence: Multiracial development and social pathways. *Child Development*, 77(5), 1298–1308.

Holmes, R. (2005). Exploring representations of children and childhood in photography and documentary: Visualizing the silence. In L. Jones, R. Holmes, and J. Powell (Eds). *Early childhood studies: A multiprofessional perspective* (pp. 164–182). Maidenhead: Open University Press.

hooks, b. (2003). Teaching community: A pedagogy of hope. New York: Routledge.

Horowitz, R.E. (1939). Racial aspects of self-identification in nursery school children. *Journal of Psychology*, 7, 91–99.

Horowitz, R.E., and E.L. Horowitz. (1938). Development of social attitudes in children. *Sociometry*, 1(3/4), 301–338.

Huddy, L. (2001). From social to political identity: A critical examination of social identity theory. *Political Psychology*, 22(1), 127–156.

Hughes, D. (2003). Correlates of African American and Latino parents' messages to children about ethnicity and race: A comparative study of racial socialization. *American Journal of Community Psychology*, 31, 15–33.

Hughes, J., R. Bigler, and S. Levy. (2007). Consequences of learning about historical racism among European American and African American children. *Child Development*, 78(6), 1689–1705.

Ingraham, C. (2004) Cinderella dreams: The allure of the lavish wedding. *Journal of Marriage and Family*, 66(4), 1069–1070.

Iwawaki, S., K. Sonoo, J. Williams, and D. Best. (1978). Color bias among young Japanese children. *Journal of Cross-Cultural Psychology*, 2, 61–73.

Iyer, A., C.W. Leach, and F.J. Crosby. (2003). White guilt and racial compensation: The benefits and limits of self-focus. *Personality and Social Psychology Bulletin*, 29(117–129).

Johnson, D. (1992). Racial preference and biculturality in biracial preschoolers. *Merrill-Palmer quarterly*, 38(2), 233–244.

Kaomea, J. (2000) A curriculum of aloha? Colonialism and tourism in Hawai'i's elementary textbooks. *Curriculum Inquiry*, 30(3), 319–344.

Katz, P.A. (1973). Stimulus predifferentiation and modification of children's racial attitudes. *Child Development*, 44(2), 232–237.

———. (1987). Developmental and social processes in ethnic attitudes and self-identification. In J.S. Phinney (Ed). *Children's ethnic socialization* (pp. 92–100). Newbury Park, CA: Sage.

———. (2003). Racists or tolerant multiculturalists? How do they begin? *American Psychologist*, 58, 897–909.

Katz, P., and S. Zalk. (1978). Modification of children's racial attitudes. *Developmental Psychology*, 14, 447–461.

Kehoe, J. (1984). Achieving cultural diversity in Canadian schools. Ontario, Canada: Vesta.

Kelly, M.D., J. Duckitt. (1995). Racial preference and self-esteem in black South African children. *South African Journal of Psychology*, 25(4), 217–223.

Kibria, N. (2000), Race, ethnic options, and ethnic binds: Identity negotiations of second-generation Chinese and Korean Americans. *Sociological Perspectives*, 43(1), 77–95.

Killen, M., N. Margie, and S. Sinno. 2006. *Morality in the context of intergroup relationships, Handbook of moral development.* New Jersey: Lawrence Erlbaum, pp. 155–185.

Kircher, M., and L. Furby. (1971). Racial preferences in young children. *Child Development*, 42(6), 2076–2078.

Kowalski, K. (2003). The emergence of ethnic and racial attitudes in preschool-aged children. *Journal of Social Psychology*, 143(6), 677–690.

Lal, S. (2002). Giving children security: Mamie Phipps Clark and the racialization of child psychology, *American Psychologist*, 57(1), 20–28.

Lasker, B. (1929). *Race attitudes and children.* New York: Henry Holt.

Latour, B. (1997) *On actor-network theory: A few clarifications.* Sourced at www.keele.ac.uk/depts/stt/stt/ant/latour.htm on August 31, 2008.

Levy, S., T. West, L. Ramirez, and J. Pachankis. (2004). Racial and ethnic prejudice among children. In J. Lau Chin (Ed). *The psychology of prejudice and discrimination: Racism in America* (pp. 37–60). Westport, CT: Greenwood.

Lord, M. (1994). *Forever Barbie: The unauthorized biography of a real doll.* New York: Avon Books.

Lubeck, S. (1994). The politics of developmentally appropriate practice. Exploring issues of culture, class and curriculum. In B.L. Mallory and R.S. New (Eds). *Diversity and developmentally appropriate practices. Challenges for early childhood education* (pp. 17–43). New York: Teachers College Press.

Lynch, J., C. Modgil, and S. Modgil. (1992). *Cultural diversity in schools.* Bristol, PA and London: Falmer Press, Taylor and Francis.

Mac Naughton, G. (2001a). "Blushes and birthday parties": Telling silences in young children's constructions of "race." *Journal of Australian Research in Early Childhood Education*, 8(1), 41–51.

———. (2001b). Silences and subtexts of immigrant and non-immigrant children. *Childhood Education*, 78(1), 30–36.

———. (2003). Eclipsing voice in research with young children. *Australian Journal of Early Childhood*, 28(1), 1–14.

———. (2004). Learning from young children about social diversity: Challenges for our equity practices in the classroom. In A. van Keulen (Ed). *Young children aren't biased, are they?! How to handle diversity in early childhood education and school* (pp. 65–76). Amsterdam: SWP.

———. (2005). *Doing Foucault in early childhood studies: Applying poststructural ideas*. London: Routledge.

Mac Naughton, G., and G. Williams. (2008). *Teaching techniques for young children*, 3rd ed. Melbourne: Pearson Education.

Mac Naughton, G., and K. Davis. (2001). Beyond "othering": Rethinking approaches to teaching young Anglo-Australian children about Indigenous Australians. *Contemporary Issues in Early Childhood*, 2(1), 83–93.

Mac Naughton, G., K. Smith, and K. Davis. (2005). A rhizoanalysis of preschool children's constructions of cultural and "racial" diversity. Paper presented at Reconceptualising early childhood education: Research, theory and practice. 13th annual conference, Wisconsin, October 16–20.

Magee, C. (2005). Forever in Kente: Ghanian Barbie and the fashioning of identity. *Social Identities*, 11(6), 589–606.

Mansfield, N. (2000). *Subjectivity. Theories of the self from Freud to Haraway*. Crows Nest, NSW: Allen and Unwin.

Mansfield, E., and J. Kehoe. (1994). A critical examination of anti-racist education. *Canadian Journal of Education*, 19(4), 418–430.

Martin, S. (2004). Dead white male heroes: Ludwig Leichhardt and Ned Kelly in Australian fiction. In J. Ryan and C. Wallace-Crabbe (Eds). *Imagining Australia: Literature and culture in the new new world* (pp. 23–52). Massachusetts: Harvard University Press.

Martino, W. (1999). "Cool boys", "party animals", "squids", and "poofters": Interrogating the dynamics and politics of adolescent masculinities in school. *British Journal of Adolescent Masculinities in School*, 20(2), 239–263.

McDonough, Y.Z. Ed. (1999) *The Barbie chronicles*. New York: Touchstone, paperback original.

McGlothlin, H., and M. Killen. (2006). Intergroup attitudes of European American children attending ethnically homogeneous schools. *Child Development*, 77(5), 1375–1386.

McGuffey, C.S., and B.L. Rich. (1999). Playing in the gender transgression zone: Race, class and hegemonic masculinity in middle childhood. *Gender and Society*, 13(5), 608–627.

McIntosh, P. (1997). White privilege and male privilege: A personal account of coming to see correspondences through work in women's studies. In R. Delgado and J. Stefancic (Eds). *Critical white studies. Looking behind the mirror* (pp. 291–299). Philadelphia: Temple University Press.

Meltzer, H. (1939). Group differences in nationality and race preferences of children *Sociometry*, 2(1), 86–105.

Moore, V. (2002). The collaborative emergence of race in children's play: A case study of two summer camps. *Social Problems*, 49(1), 58–78.

Moran, S. (2004) White lives in focus: Connecting social praxis, subjectivity and privilege. *Borderlands e-journal*, 3(2).

Moreton-Robinson, A. (2000). *Talkin' up to the white woman: Aboriginal women and feminism.* St Lucia, Qld: University of Queensland Press.

———. (2004). Whiteness, epistemology and Indigenous representation. In A. Moreton-Robinson (Ed). *Whitening race. Essays in social and cultural criticism* (pp. 75–88) Canberra: Aboriginal Studies Press.

Morland, K.J. (1963). Racial self-identification: A study of nursery school children. *The American Catholic Sociological Review*, 24(3), 231–242.

———. (1969). Race awareness among American and Hong Kong Chinese children. *The American Journal of Sociology*, 75(3), 360–374.

Morrison, T. (1992). *Playing in the dark: Whiteness and the literary imagination.* New York: Harvard University Press.

Murrie, L. (1998). The Australian legend: Writing Australian masculinity/writing "Australian" masculine. *Journal of Australian Studies*, 56, 68–78.

N/A. (1929). Review: [untitled]. *The Journal of Negro History*, 14(3), 352–353.

Nelson, A.K. (1929). Review: [untitled]. *Educational Research Bulletin*, 8(15), 351–352.

Nesdale, A. (1987). *Ethnic stereotypes and children.* Paper No. 57. Clearinghouse on Migration Issues.

Newton, J. (1998). White guys. *Feminist Studies*, 24(3), 572–598.

Nicoll, F. (2004). Reconciliation in and out of perspective: White knowing, seeing, curating and being at home in and against Indigenous sovereignty. In A. Moreton-Robinson (Ed). *Whitening race* (pp. 17–31). Canberra: Aboriginal Studies Press.

Nilsen, B.A. (1997). *Week-by-week: Plans for observing and recording young children.* New York: Delmar.

Norton, M.I., S.R. Sommers, E.P. Apfelbaum, N. Pura, and D. Ariely. (2006). Colorblindness and interracial interaction: Playing the political correctness game. *Psychological Science*, 17, 949–953.

O'Loughlin, M. (2001). The development of subjectivity in young children: Some theoretical and pedagogical considerations. *Contemporary Issues in Early Childhood*, 2(1), 49–65.

Palmer, G. (1990). Preschool children and race: An Australian study. *Australian Journal of Early Childhood*, 15(2), 3–8.

Park, R.E. (1928). The bases of race prejudice. *Annals of the American Academy of Political and Social Science*, 140, 11–20.

Pease, B. (2004). Decentring white men: Critical reflections on masculinity and white studies. In A. Moreton-Robinson (Ed). *Whitening race. Essays in social and cultural criticism* (pp. 119–130). Canberra: Aboriginal Studies Press.

Petrecca, L. (2005). Disney hopes fairies will fly into girls' hearts. *USA Today*, August 25. Sourced at asp.usatoday.com/community/utils/idmap/13064889.story on August 31, 2008.

Piaget, J. (1954). *Intelligence and affectivity: Their relationship during child development.* Palo Alto, CA: Annual Review.

———. (1968). *Genetic epistemology.* New York: Columbia University Press.

Pyke, K., and D. Johnson. (2003). Asian American women and racialized femininities: "Doing" gender across cultural worlds. *Gender and Society*, 17(1), 33–53.

Quintana, S.M., and C. McKown. (2008). Social identity development and children's ethnic attitudes in Australia. *Handbook of race, racism, and the developing child*, Hoboken, NJ: John Wiley, pp. 313–338.

Ramsey, P. (1991). The salience of "race" in young children growing up in an all-white community. *Journal of Educational Psychology*, 83(1), 28–34.

Reinhardt, J. (1928). Students and race feeling. *Survey*, 61, 239–240.

Renold, E. (2004). "Other" boys: Negotiating non-hegemonic masculinities in primary school. *Gender and Education*, 16(2), 247–266.

Reuter, E.B. (1929). Review: [untitled]. *The American Journal of Sociology*, 35(3), 491–493.

Richardson, Brian. Ed. (2005). *Tell it like it is: How our schools fail black children.* London: Bookmarks Publications and Stoke on Trent: Trentham Books.

Roediger, D. (1994). *Towards the abolition of whiteness: Essays on race, politics, and working class history.* New York: Verso.

Rolfe, S.A. (2001). Direct observation. In G. Mac Naughton, S.A. Rolfe, and I. Siraj-Blatchford (Eds). *Doing early childhood research. International perspectives on theory and practice* (pp. 224–239). NSW: Allen and Unwin.

Sachs. J. (1999). Using teacher research as a basis for professional renewal. *Journal of In-Service Education*, 25(1), 39–53.

Semaj, L. (1981) The development of racial-classification abilities. *The Journal of Negro Education*, 50(1), 41–47.

Silin, J. (1999). Speaking up for silence. *Australian Journal of Early Childhood*, 4(24), 41–45.

Simon, R.J. (1974). An assessment of racial awareness, preference, and self identity among white and adopted non-white children. *Social Problems*, 22(1), 43–57.

Singleton, L.C., and S.R. Asher. (1979). Racial integration and children's peer preferences: An investigation of developmental and cohort differences. *Child Development*, 50(4), 936–941.

Skattebol, J. (2005). Insider/outsider belongings: Traversing the borders of whiteness in early childhood. *Contemporary Issues in Early Childhood*, 6(2), 189–203.

Spencer, M.B., and C. Markstrom-Adams. (1990). Identity processes among racial and ethnic minority children in America. *Child Development*, 61(2), 290–310.

St. John, N., and R. Lewis. (1975). Race and the social structure of the elementary classroom. *Sociology of Education*, 48(Summer), 346–368.

Stevenson, H., and E. Stewart. (1958). A developmental study of racial awareness in young children. *Child Development*, 29(3), 399–409.

Targowska, A. (2001). Exploring young children's "racial" attitudes in an Australian context—the link between research and practice. Paper presented to the Australian Association of Research in Education annual conference.

Tascon, S. (2002). Refugees and asylum-seekers in Australia: Border-crossers of the postcolonial imaginary. *Australian Journal of Human Rights*, 8(1). Sourced at http://www.austlii.edu.au/au/journals/AJHR/2002/9.html on August 29, 2008.

Taylor, A. (2004). Playing with difference: The cultural politics of childhood belonging. *The International Journal of Diversity in Organisations, Communities and Nations*, 7(3), 143–150.

———. (2005). Situating whiteness critique in Australian early childhood: The cultural politics of "Aussie" kids in the sand pit. *International Journal of Equity and Innovation in Early Childhood*, 3(1), 5–17.

Thoma, P. (2007). *Of beauty pageants and Barbie: Theorizing consumption in Asian American transnational feminism.* Sourced at http://social.chass.ncsu.edu/wyrick/DEBCLASS/barbie.htm#note34 on August 31, 2008.

Thompson, A. (1999). Colortalk: Whiteness and off white. *Educational Studies: A Journal of the American Education Studies Association*, 30(2), 141–160.

Trew, K. (2004). Children and socio-cultural divisions in Northern Ireland. *Journal of Social Issues*, 60(3), 507–522.

Turner, P., J. Gervai, and R. Hinde. (1993). Gender typing in young children: Preferences behaviours and cultural differences. *British Journal of Developmental Psychology*, 11, 323–342.

Van Ausdale, D. (1996). *Preschool children's use of race and ethnicity in everyday interaction.* Unpublished PhD thesis. University of Florida.

———. (2004) Children. In E. Cashmore (Ed). *Encyclopedia of race and ethnic studies* (pp. 78–81). London and New York: Taylor and Frances.

Van Ausdale, D., and J.R. Feagin. (1996). Using racial and ethnic concepts: The critical case of very young children. *American Sociological Review*, 61(5), 779–793.

———. (2001). *The first R: How children learn race/racism.* Maryland: Rowman and Littlefield.

Van Driel, B. (2004). *Confronting Islamophobia in educational practice.* London: Trentham.

Varney, W. (1998). Barbie Australis: The commercial reinvention of national culture. *Social Identities*, 4(2), 161–176.

Vaughan, G.M. (1986). Social change and racial identity: Issues in the use of picture and doll measures. *Australian Journal of Psychology*, 38(3), 359–370.

Vora, P., and E.M. England. (2000). Children's social categories and the salience of race. Research report sourced at http://www.eric.ed.gov/ERICWebPortal/custom/portlets/recordDetails/detailmini.jsp?_nfpb=true&_&ERICExtSearch_SearchValue_0=ED441563&ERICExtSearch_SearchType_0=no&accno=ED441563 on June 21, 2008.

Wadham, B. (2004). Differentiating whiteness: White Australia, white masculinities and Aboriginal reconciliation. In A. Moreton-Robinson (Ed). *Whitening race. Essays in social and cultural criticism* (pp. 192–207) Canberra: Aboriginal Studies Press.

Wang, A. (2000). Asian and white boys' competing discourses about masculinity: Implications for secondary education. *Canadian Journal of Education*, 25(2), 113–125.

Wen, P. (2000). Three faces of Barbie a snub to Asians. *The Boston Globe*, May 27.

Whitesell, N.R., C.M. Mitchel, C.E. Daufman, and P. Spicer. (2006). Developmental trajectories of personal and collective self-concept among American Indian adolescents. *Child Development*, 77(5), 1487–1503.

Yip, T., E.K. Seaton, and R.M. Sellers. (2006). African American racial identity across the lifespan: Identity status, identity content, and depressive symptoms. *Child Development*, 77(5), 1504–1517.

Yoon, C. (1994). A Doll of our own. *A. Magazine* (April/May), 28–32.

Young, D. (1927). Some effects of a course in American race problems on race prejudice of 450 undergraduates of the University of Pennsylvania. *Journal of Abnormal and Social Psychology*, 22, 235–242.

Young, R.J.C. (1998). *Postcolonialism: An historical introduction.* Oxford: Blackwell.

Zinser, O., M.C. Rich, and R.C. Bailey. (1981). Sharing behavior and racial preference in children. *Motivation and Emotion*, 5(2), 179–187.

PART II

Exploring the Politics of Adults' Racialized Identities

CHAPTER 6

(Un)masking Cultural Identities: Challenges for White Early Childhood Educators

Karina Davis

Introduction

It has been argued within the literature on critical race theories that individuals and groups who are white do not often see themselves as being connected to a cultural or racial group. For white individuals and communities, the cultural basis of their identities has been silent and unnamed, and largely remains so, even while their identities are discursively constructed from within particular social, historical, and geopolitical times (Delgado and Stefancic, 1997; Frankenberg, 1993; Moreton-Robinson, 2004). It is undesirable and problematic to collapse the diversity and multiplicity that exists in white communities, indeed any community, into one static and concrete category, and this work requires the explorations of historical, geopolitical, and social constructions of the specifics of any context and country. With these tensions in mind however, it is also useful to begin naming white to ensure that white individuals and groups begin to locate themselves as just another cultural group. This is useful, as it provides entry-points and opportunities to begin to trouble and problematize the silent and unacknowledged judgment and categorization that occurs for "others" against the unnamed and unspoken white norm. In Australia specifically, the necessity of (un)masking white identities in efforts to decenter white and accord it just another place in a diverse racial and ethnic mix is particularly relevant. As a country, Australia has struggled to reconcile its Indigenous history and present with its colonial occupation, and continues to struggle in achieving its multicultural agenda and policies of building respectful and socially just realities for many of its citizens (Moran, 2005; Neumann, 2004). The education system in Australia is implicated and rearticulates this struggle and there are calls for these systems to lead the work in exploring and shaping racial and cultural justice and equity for all children and families (McLaren, 1997; Mac Naughton, 2005).

In order for educative work to explore cultural diversity and "race" and construct relationships with children, families, and colleagues that are equitable and socially just, educators, especially white educators, must work to see their own cultural identities as being part of the culturally diverse mix. In this, white educators must see the object for foci and analysis not in the exotic "other" but instead work to locate their own white cultural identities, recognize their identities as "race"d, and see themselves as intimately connected with issues surrounding "race." This analysis must work to uncover how white educators' silencing of their own cultural identities also works to silence their connection with dominance and racial and cultural oppressions and inequities in order for pedagogy and relationships to be constructed from more respectful and socially just understandings.

What follows in this chapter is a mapping of what I would loosely name a "process" in this work toward (un)masking white identities in order to trouble culturally based assumptions, stereotypes, and silences in efforts to explore educative work that is socially just. It draws from data from a small qualitative research project (refer appendix for further details) conducted in Melbourne, Australia, which focused on exploring the ways in which a group of eight early childhood educators worked to name white cultural and racial identities, explorations of what became important to uncover, and a discussion of the possibilities and questions that were raised for the group from these explorations. In exploring this, I will use extracts from the project's research meetings, using pseudonyms for the research participants.

White Identities as Masked

> Why have white, you know. Why have the color? Why does it have to be colored? (Elizabeth).

Elizabeth's questioning of why it is necessary to name white is indicative of the silencing and unfamiliarity with naming and locating white as a cultural and racial identity. For many white individuals, white is not only masked but its (and subsequently their own) connections to "race" remain invisible too. The color of white skin is often not spoken of as a marker of identity, especially not a cultural or racial identity, while the color of black or nonwhite individuals or communities is and remains a powerful marker of other identities for whites. Black and nonwhite others are categorized, discussed, and labeled by whites in accordance with physical and/or audible racial, cultural, or linguistic markers. However, this labeling and cultural and racial categorization of white individuals and communities from within is largely absent (Cannella, 1997; Fine et al., 1997; Morris and Cowlishaw, 1997).

This absence of cultural and racial categorization from within is partly a result of white cultures having cemented their understandings of themselves and their worlds as universal for all humanity (Frankenberg, 1993; Moreton-Robinson, 2000) and through these processes masking the cultural basis of these understandings. White understandings of, for example, growth, development, science, creation, appropriate and inappropriate behaviors are seen as desirable for all people and are perceived as the most advanced and true ways of looking at and interpreting the world (Cannella, 1997). Within white discourses, these truths are seen as universal and it is not recognized that they have been, and continue to be, developed along cultural lines within specific historical, social, and political circumstances. Through these processes, and processes like othering, black and nonwhite cultures are seen as being a culture or "race" while white remains unmarked and unnamed, a silent and powerful norm in which all others are compared, normalized, and expected to conform (Howard, 1999; Ladson-Billings and Tate, 1995).

With these discursive constructions of white in circulation, it is not difficult to see why Elizabeth struggles with seeing how and why she would seek to name white. Naming a white racial and cultural identity has had little relevance in her life, and her own connections to her white Australian culture have existed at times as a cultural void, especially when compared to the perceived depth of cultural histories and traditions of "others." As she explains:

> I sort of identify more with my Irish background than my Australian background...it's much easier for me if people ask "Oh, what's your background?" It's much easier to go into the Irish side. (Elizabeth)

The dichotomy between an unnamed and lacking-in-culture white and a culture-rich "other" is one of the implications of white universalizing discourses. As white understandings, traditions, and histories are simultaneously normalized, universalized, and unnamed, histories and traditions that sit outside these norms are seen as exotic and/or deviant. "Others" are celebrated and accepted for cultural and racial differences that provide curios for white communities but are sanctioned and judged for differences that are perceived as at-odds with or threatening to white sensibilities. In these ways, the exotic and deviant "other" is seen only in opposition to normalized and cultureless white, while the white individual and community remains the arbitrator for what is accepted. Within these discursive processes of white, having and belonging to a culture is seen as connected only to the "other" and only marked by cultural practices that oppose the white norm (Gallini, 1996; Hage, 2000). For Elizabeth, this dichotomy is marked by her ease in discussing and naming her identity as connected to her othered Irish background (while it is acknowledged that this remains an Anglo-Irish identity) and the difficulty she has in naming and articulating her "Australian" identity.

Constructing a Paradox: White Identities as the Australian National Image

> People always ask if I say I'm Australian "...but no, where do you come from?" (Anna)

In an Australian context, the representational image of an Australian is one of white-Anglo founded on colonial and white histories (Elder, 2005; Hage, 2000; Kelen, 2005). This image is often reinforced to Anna, as she explains above, when she attempts to claim an Australian identity for herself. The Australian national image is based on colonial stories and romanticized mythology of the white-Anglo migrant battling against the geography and climate of a remote and difficult landscape. This white Australian as the national image has been further reinforced across time through migration policies and practices that continue today and that privilege white and exclude "others." Although today Australian politics names multiculturalism as a national policy and proudly acknowledges the depth and importance of the nation's cultural diversity, it is "others" that are positioned as the culturally diverse against an unacknowledged and unnamed white background and identity (Elder, 2005; Moran, 2005; Russell, 2005). While in these national conversations white is masked and remains unacknowledged from within, this white is simultaneously marked as an essential component of an Australian national image. This constructs a paradox in which white is both masked and marked within white communities—masked and used to deny "race" and marked to embrace an exclusionary national identity. In this Australian context, Anna's statement above stands as a reminder that unnamed white pervades and carries a claim to an exclusionary national identity not available to all.

These discursive constructions of white national identities and belonging were given further voice in a later research group conversation. This conversation was one that dominated one particular research group meeting and the topic was often reintroduced in later meetings during the research process. The extract below draws from an exchange between Anna and Lisa, as Lisa discusses how she often asks others "Where are you from" as a way of initiating a conversation:

> *Lisa*: ...generally if you ask "Where are you from" you are trying to establish a point of conversation. So we innocently ask questions. Knowing for some people it can be offensive...but we don't set out to be offensive. I think you've just got to go with your gut instinct...it's how we get a foot in the door to connect with people.
>
> *Anna*: But it is that fact if you saw a blonde-haired blue-eyed girl you would not ask "Where do you come from?"
>
> *Lisa*: (pause)...But are we too precious about it sometimes?
>
> *Anna*: But I think if you're not the one getting asked you're probably less precious about it, if that's the word you want to use.

This extract illustrates how white discursive meanings and values embedded in the national image of a white Australian are not completely invisible or transparent for white individuals and groups. As Lisa explains her point, she draws from understandings and constructions of her own white identity and links this to her white community—she discusses that "we" don't mean to be offensive, that "we" need to get a foot in the door with others. But throughout the conversation she works to maintain the image of a white identity as the identity that belongs to the nation. This is achieved through a process of avoiding naming white, while fixing the focus on the "other." This other is known in contrast to the "we" Lisa refers to—a normalized and white national identity. When Anna challenges this normalized and exclusionary white national identity, Lisa works to counter this discussion through silencing the racial basis to this exclusion and as such, disconnecting issues of "race" from the national identity. In her attempt to trivialize the exclusion ("But are we too precious about it...?") Lisa moves to deny "race" and racism through locating exclusion as the perception of an overly sensitive (othered) individual. In this trivializing she also works to mask the discriminatory implications of inequitable cultural and racial positionings. In responding to this, Anna challenges this trivialization and works to unmask and name the white national identity as one constructed from discriminatory and exclusionary racially based understandings.

Exclusions and inclusions based on racial and cultural categorizations, like the one in the conversation above, are a part of white discourses and work to construct exclusions of others as a result. While these exclusions and inclusions might not always be seen by white individuals as having sinister or discriminatory intent (as discussed by Lisa above in the comment "but we don't set out to be offensive"), they are often based on understandings of white as the universalized norm. At this point it is important to acknowledge that within white discourses, cultural and racial identities are hierarchically organized and different identities are accorded different values and places in the hierarchy. This hierarchical location is fluid and is also intersected by other aspects including gender, class, and language that result in different and varied experiences of cultural and racial identities for many (Frankenberg, 1993). Currently in Australia, in regards to racial and cultural hierarchies, white identities are routinely placed at the top, with other cultural and racial identities organized beneath (Elder, 2005; Hage, 2000; Moreton-Robinson, 2004). As a result of this, while white individuals face many disadvantages and inequities themselves around issues of gender, class, and the like, they are positioned powerfully in terms of cultural and racial identity and subsequently accorded a connection and place within the national image. In the case of the discussion between Lisa and Anna, the existence of a racial hierarchy is illustrated in the placing of white identity as the unnamed and powerful norm. Anna, and other individuals and groups marked as nonwhite, are routinely excluded from any assumption of national belonging through exclusionary acts that are seen as unproblematic from within white discourses. It is only possible to ask an individual where they come from if you have an understanding that they do not fit within a particular construction of nation and identity. Further, that Lisa used discursive practices to reinforce exclusion in ways she was not aware of and/or failed to see as problematic highlights her place of power within a hierarchy that does not present challenges for her around "race" and belonging.

Possibilities in the White Discursive Paradox

So what does it mean to position white as a marker of identity but a marker that is largely unnamed and unspoken? How is it possible to both position a white identity as unacknowledged and unarticulated but to also rely on this identity to mark both yourself and others and form decisions and actions around inclusions and exclusions? White discourses working to simultaneously silence connections and discussions of "race" and culture while relying on the identification of white as the marker of inclusion/exclusion makes mapping the workings of white discourses problematic and challenging. It is in exploring these tensions and contradictions within this paradox with white individuals, however, where some of the possibilities for disrupting and deconstructing white discourses exist. In education specifically, white educators can work these tensions and contradictions to locate and explore their own white identities as raced and cultured, provide spaces where possibilities are created for seeing their own intimate connections with and to "race," and work to engage with children and families in ways that are more socially just and equitable. The next section of this chapter explores how meeting and remeeting white for the research group members made these sorts of explorations possible.

Meeting and Remeeting White

In the research group, possibilities for meeting and remeeting white came from exploring the tensions and contradictions in white discursive constructions of white identities. While always planned as part of the research process, meeting and exploring white constructions of identities became an increasingly important focus of the research group meetings and activities. It became clear to all group members that this meeting of white identities provided possibilities for naming and articulating either their own white identities or provided space and language to uncover and critique inequitable white processes. This also provided spaces for seeing the implications of white identity constructions personally and within pedagogy and educative practices more clearly.

As has been discussed previously, white is slippery and difficult to name and articulate even as it is used as a basis for white individuals and communities to construct and understand identities. This slipperiness and difficulty in naming white was no less apparent for the research group members. In efforts to begin to work on disrupting white identities, the research project overtly expressed the aim of working to understand the role white played in educative work for and with cultural diversity. The research group members where engaged in conversations aimed at naming white from the beginning of the research process. This engagement had different meanings for each of the research members and the research process was constructed around attempting to layer disruptions at a point and in a direction or focus they each identified as desirable and necessary for them. These disruptions had a variety of effects, three of which—naming white, meeting the other, and critical engagement with white identities—are discussed in the following sections.

Naming White

> When I think about it (my white cultural identity) in a more critical way...I think it is just a lack of recognition. But before that, I probably would have just felt it was a "lack of." And I would never even...noticed it as a culture. (Elizabeth)

> The reading on white has been a real eye-opener for me. Like WOW. I just don't think that people know this...we don't know. (Sue)

For many white individuals, the act of naming and locating white as a cultural group is in itself a disruptive act that provides space for shifts and critical engagement (Berlak and Moyenda, 2007). For Sue and Elizabeth, engaging with text and conversations that named white began a process in which they began to see white as another cultural group and consider the implications of dominant, normalized, and unnamed white identities. For Sue, who had requested readings that could explain to her "what this white stuff actually means," the reading of these white texts, in her words, "opened her eyes." Elizabeth was also clearly able to articulate that before this exposure to the naming of white discourses and identities, she had never named herself culturally and that it was the active naming that allowed her to see the lack of recognition given to white identities.

Specifically for Sue, this initial disruption provided a catalyst and space to begin to reflect on what this lack of naming and acknowledgment of white has meant for her engagements with cultural and racial identities over her life. In the extract below she discusses how this provided a powerful shift in her understandings of both herself and others in ways that opened up multiple possibilities and disrupted and deconstructed stereotypical understandings of others.

> I failed to really grasp the concept…that white people have a cultural background like all others and are not the group to be compared against. Also I have really struggled to understand until very recently that there is really no stereotype within any cultural group and that all people live their lives in different ways. (Sue)

Meeting the Other

Another strategy that was used within the research group to disrupt white discourses and identity constructions was to introduce group members to the voices of others. In this, research group members where able to explore how white and white identities have been constructed and framed by those who are black and nonwhite. The research group members were also able to hear and engage with histories and experiences of black and nonwhite communities that construct problematic images of white identities—as colonizers, as abusers of power, as discriminatory. This is especially important in the work of the research aimed at disrupting white, as the voices of others have historically been silenced, especially in white colonized countries like Australia, and continue to be marginalized today (Hage, 2000). White voices largely construct and filter national histories, politics, entertainment, and media. The research group members did not have histories themselves in engaging with voices and texts that reached outside of these white constructions; and for some like Sue, this was the first time they saw themselves as visible and culturally located.

> It has been very confronting for me to face the possibility that the things I have been saying may have caused others to feel uncomfortable or to reiterate feelings of being made to feel different. (Sue)

Using the voices of others to highlight white as a cultural and racial location is an important part of exploring white discursive constructions of identities. The meeting of othered voices for Sue made it possible for her to see what the implications are of an unnamed and normalized white that works to position others as exotic and racialized. In white constructions these understandings of the other are often based on stereotypical, simplistic and derogatory understandings and images and Sue highlights the possibilities for disruptions that meeting the other can produce.

Critical Engagement With Educator Identities

> But it (the experience of critical engagement) was enough, definitely, for me to be completely changed in how I looked at everything. Everything! (Sonia)

One other consequence of the work on disrupting and deconstructing constructions of white identities for the research group members highlighted by Sonia was the increased critical engagement they felt with their educative roles as a whole. All research group members discussed that critical engagement with theory and with texts, and the questioning of each other, of themselves and their pedagogy that resulted, had flow-on effects for how they considered their role as educators and their engagement with education overall. Sonia felt empowered by what she saw as possible once critical engagements refocused her gaze. As for Elizabeth, she explains:

> I guess it (the effects of critical engagement) was . . . like I would have always thought of myself as a . . . sort of a fair . . . person that was caring about other people and things but then . . . I started to think about and I realized . . . the, stereotypes that were imbedded in my thinking and . . . the value that was put on particular sort of practices and behaviors and . . . even down to like, sitting on a train or something and I'll sort of make judgment calls about the people around me. You know, whereas, now I'll do that a lot less. (Elizabeth)

These engagements and realizations that arose for the research group members, like Elizabeth and Sonia, were not always easy ones to work and live through, and required a commitment to the process and belief in the work they were doing. Within this though what appeared to be sustaining was a strong belief that this engagement, and the disruptions connected to it, was making them fairer and more respectful people, and as a consequence, more socially just educators who could take action in their practices in ways that made lives fairer and more equitable for all. As Sue explains:

> It wasn't until I reflected on (my work with a nonwhite young boy) . . . that I realized I was making judgments and I was making my decisions based on my culture and what I sort of knew and not taking

> into consideration theirs, so...that experience really nearly knocked me off my pedestal. I struggled to get through that year because it was so stressful. And it was almost like it was the start of something and now I can actually process that and stop feeling guilty but certainly be a lot more aware of how I could've done it and how, in another circumstance, I certainly will. (Sue)

What Now? Implications for Antiracist Pedagogies

Through the research meetings it was highlighted that critical engagement with white became an accepted and necessary part of pedagogical and educative practices. With this in mind though, what can we draw from and learn from the experiences of this research group—for local and international contexts? What was disrupted and deconstructed and what changed as a result? How do we map these changes? What are they useful for? What can we take to an international audience? Can changes in practices result?

In the extract below, Lisa explores the challenges involved in working to see and name the complexities involved in shifting and reassessing white identity constructions in our daily lives. While this reassessment occurred in this project in an Australian context, the shifts and reassessments are relevant to international contexts where white discourses work to dominate and discriminate, as they can highlight how local explorations of how white works to discriminate can begin. Lisa highlights her explorations in a conversation that recounted an experience of a physical confrontation between two groups of adolescent men. In this extract, she highlights the important struggles white individuals and educators must engage with in order to trouble racist practices. Lisa illustrates how she works to name white and disrupt the image of the "other" in ways that recognize white identity constructions as simplistic and problematic and seeks to find ways that construct this episode in more complex and fairer ways.

> I felt so angry towards the aboriginal boys and found that difficult due to the fact that I kept thinking there may be such complex background issues...History seems to have a lot to answer for and when you know some things of this, it is perhaps, tied up into such complexity that will continue to impact on us as Anglos and Indigenous peoples for a long time to come. I could not separate that these were...two boys threatening two other boys, (from) the skin factor and our past history along with the fact that the boys live at an

> Aboriginal settlement still in existence. Maybe this is why it is so uncomfortable and we do get confronted in our teaching roles. (Lisa)

Lisa's acknowledgment that there are no easy answers in dealing with cultural and racial differences, her willingness to engage in disrupting the dominant white construction of Indigenous Australian people and her work to complexify the issues around a confrontation are what are sometimes made possible when we disrupt white. While Lisa's response and reflections did not completely disassociate her from a colonial white identity (and it could be argued that this disassociation is impossible anyway), her willingness to disrupt these constructions is encouraging and provides us with hope for the work against white normalization and dominance in any context. In another research extract (below), Sue also outlines how her critical engagements with identity have made it possible to see and name inequities and categorizations and highlights how she now works to disrupt these in institutional processes.

> I received a form from the local council at work a few weeks ago asking me to identify the families that did not originate from Australia. I found myself unable to answer as there was no definition of what they considered as not originating from Australia. I have several children who have one parent that was born outside Australia but does that constitute "not originating from Australia" and ultimately even the Indigenous Australians originally came from somewhere else! Furthermore I was troubled as to why they would need to know this information. So I (and the president of the committee who was experiencing the same struggle) responded to the question by stating that we were not prepared to answer the question. (Sue)

In these extracts, both Lisa and Sue highlight how critical engagement with white discourses and identity constructions can provide opportunities for changes in discursive practices. For both these white women there have been significant shifts in how they engage in their social worlds, including how they engage as educators. These shifts are optimistic signs of some of the possibilities that are created when we work to name and disrupt white and how white educators can deconstruct their own identities in ways that support their work for social justice. The processes of naming white, meeting the other, and critically engaging with educator identities can be explored within many contexts and nation-states and requires consideration of the historical, geopolitical, and social contexts. With this knowledge, educators can begin to identify who the "other" might be in their context, and work to unmask, confront, and challenge the local specificities of white dominance and oppression that exists.

The research group meetings proved powerful in uncovering previously silenced and/or unnoticed moments of identity constructions based on white discourses; and the critical engagements provided possibilities for greater shifts in perspectives and practices. While these disruptions and engagements occurred in this particular context, this work is far from finished for all of us within the research group. This meeting of white was (and is) never complete, as white discursive practices are always in operation and working to draw white individuals and communities back into dominant constructions of identity that rely on universalizing and normalizing white. This can be seen within the processes and conversations of the research group as even while white identities were uncovered and critiqued in one place at one time, other silences and inequities existed and were rearticulated in another space and time. In this way, for white individuals and communities, their work in exploring and locating white is always incomplete and requires continual reflection and engagement.

If we reflect on the call of critical race scholars working in education, we can see shifts in the practices of the research participants in ways that heed some of their calls. Critical race scholars working in education have asked white educators to work to recognize, name, and disrupt the unnamed white constructions of identities that shape and constrain the lives of educators, families, and children within educational institutions. They argue that work focused on this would make educative spaces fairer and more equitable for all, and reconstruct the inequitable relationships between white educators and black and nonwhite colleagues, families, and children in ways that open access and possibilities. So have we worked toward this? Within the research context, have we begun to achieve this disruption of identities in ways that supports changes in pedagogy and practice? Are the shifts in seeing and acknowledging white and working in ways that disrupts white constructions of cultural and racial identities that have occurred for the research members shifts that make the lives of children in education fairer? For the research group members, the shifts and disruptions that they have encountered have made it possible for them to see, name, and be concerned about how white structures their lives and all have worked to change the ways this influences their lives. As a consequence of these shifts in white identity constructions then, educative practices must surely follow.

CHAPTER 7

On "Race" and Resistance: Transforming Racialized Identities—A Personal Journey

Merlyne Cruz

Deconstructing "Race"

I have been thinking about "race" for a few years now. My recent visit to my birth country (my second after being away for twenty-two years) triggered more curiosity on the subject matter. What amazed me most on this trip was seeing a huge range of skin-lightening products dominating the Philippine skin-care market. Sitting on the shelves of multiple alleys in various stores were a wide variety of skin-lightening products. Lining the streets and highways of urban Manila are endless rows of building-sized billboards of local artists (even politicians!) endorsing skin-whitening products.

Things have simultaneously changed and not changed, I thought. Fair/white skin is still an aspiration common to many of us. What's different this time is that the preoccupation has become deeper, completely conspicuous, yet widely accepted. A good, healthy brown never attracts admiration among us, observed Manuel Quezon Jr. over four decades ago in his article in *Philippine Graphic*. Quezon (1966) contends that Filipinos connect features and coloring with social and economic position and possibilities. There are false assumptions that a dark-skinned Filipino is uneducated, works lowly jobs, and comes from a lower socioeconomic class; the fair skinned is supposedly educated and has, or is capable of holding, white-collar jobs.

Quezon concludes that, as a people, we link intelligence and/or ability with "race": the darker skinned are seen as less intelligent compared to the fair skinned. (The effect of this is privileging white.) He then encourages Filipinos to believe that "as a people we have no inborn inferiority" and that "we have the same inborn potentialities as other people."

Not long after my trip, I watched a launching of a Filipino cultural event on television. The leader of the ceremony, after delivering his speech, beckoned the audience to chant "Mabuhay ang lahing kayumanggi!" (In English, it translates: "Long live the brown 'race'!"). The phrase is an invitation for Filipinos to respect his/her "race"—the brown "race"—and take pride in one's Filipino identity. As I pondered on this phrase, thinking about the conflicting/contradictory perspectives I learnt about "race," I couldn't help but ask many questions: What is "race"? Is there such a thing as "race"? How should I perceive "race"?

In his book, *Wretched of the Earth*, Frantz Fanon (1963) argues, "What divides the world is first and foremost...what race one belongs to." Nazneen Kane (2007) maintains that for Fanon:

> Race is not a biological trait but rather a historically constructed phenomenon and culturally mediated artifact...Human comparison is what bestows individuals with their sense of inferiority and superiority, in effect with their sense of racist ideologies based on racial signification...it is not only difference that is historically constructed but also the social signifiers, the system of (de) valuation associated with that difference. (p. 356)

The neat division of the human population into distinct "race"s, which has favored the interests, worldview, and structural location of people descended from Northwestern Europeans and their allies, has been proven to be scientifically unfounded (Miles, 1982), yet we continue to be shaped by this understanding. Cultural anthropologist Faye Harrison (2002) describes "race" as a significant marker of positioning and power and maintains that legal codifications of racial distinctions have historically played a major role in "naturalizing" the markedly unequal and unjust distribution of power and prestige over past centuries. From the vantage points of the subordinate segments of racially stratified societies, "race," a social construction (Frankenberg, 1993), is experienced as a form of symbolic as well as materialized violence.

The Existence of the Category of "Race" Is Viewed as the Outcome of Racialization

Racialization, an ideological process (Omi and Winant, 1986), is the medium through which "race" thinking operates and the process by which meaning and valuation of white are derived from the demeaning and devaluation of others (Martinot, 2003). Over the years, I have come to understand that "race" is not only about color. It is about wealth and privilege and is centered on the idea of supremacy and subjugation of peoples (Armstrong and Ng, 2005). I have also learned that race is "a purely imaginary social fabrication for the purpose of establishing a hierarchy among people" (Armstrong and Ng, 2006, p. 35).

David Goldberg (1993) traces the beginnings of the concept of "race" in colonialism. In his book *Racist Culture: Philosophy and the Politics of Meaning*, Goldberg explains:

> As West Africa was explored, conquered and its peoples enslaved by the Spanish and Portuguese, and then as the New World was created, subjugated and plundered, the concept of "race" became explicitly and consciously applied. (p. 21)
>
> In the 16th century, domination of inferior by superior was considered a natural condition...By the 17th century, racial modes of identity and identification have been embraced. (p. 25)

Franz Fanon (1967) holds European colonialism responsible for "racialization of thought": "colonialism did not dream of wasting its time in denying the existence of one national culture after another" (Fanon, 1967, p. 171). Molefi Asante (2006) defines colonialism as

> (seeking) to impose the will of one people on another and to use the resources of the imposed people for the benefit of the imposer. Nothing is sacred in such a system as it powers its way toward the extinction of the wills of the imposed upon with one objective in mind: the ultimate subjection of the will to resist. An effective system of colonialism reduces the imposed upon to a shell of a human who is incapable of thinking in a subjective way of his or her own interest. In everything the person becomes like the imposer. (p. ix)

Unsal (2006) elaborates on the social basis of this mindset by quoting Fanon's (1963) arguments:

> racial and colonial forms of oppression dehumanize people by generating a self-understanding among the oppressed that they are outside

> the scope of humanity... this process either induces the relinquishment of personal autonomy accompanied by the adoption of the view of self that was enacted upon the oppressed by the oppressor, or provokes violence in response to dreading the loss of a sense of interconnectedness with others. (Unsal, 2006, p. 66)

Albert Memmi likewise believes that racism is underpinned by colonialism. For Memmi, racism is the "profitable utilization of difference" (2000, p. 14). He argues that the struggle against racism is a fight for the social health of the collective. In this struggle, differences are not denied; instead they are recognized, embraced, and respected as such (Memmi, 2000, p. 155).

David Theo Goldberg (1990) in his work *Racist Culture* pointed out that today "race" is irrelevant yet everything is about "race." Deconstructing the idea of "race" is new to me although, as a Filipina (now living in Australia as a migrant for the past twenty-two years), I have always been somewhat consciously aware of "race" as a category. What I was not cognizant of until a few years ago was how "race," as an organizing principle of social relationships, could actually shape my identity as an individual actor at the micro level and shape all spheres of my social life at the macro level (Omi and Winant, 1994). I realize later in my life that "race" has remained/could remain such a powerful, common-sense idea that one could actually be imprisoned by it (Hall, 1997).

Racism and Colonialism: A Personal Journey

Colonial discourses of "race" and racism occupied a significant space in the cultural realm I have inherited. My native land, the Phillippines, had been ravaged by a long history of colonization—first Spain for 350 years then by the United States, for nearly half a century. Our national hero Jose Rizal wrote about the consequences of the long period of Spanish rule on the Filipino mind:

> (they) gradually lost their ancient traditions, their recollections—they forgot their writings, their songs, their poetry, their laws, in order to learn by heart other doctrines which they did not understand; other ethics, other tastes, different from those inspired in their "race" by their climate and their way of thinking. Then there was a falling-off, they were lowered in their own eyes, they became ashamed of what was distinctly their own... their spirit was broken and they acquiesced. (Rizal, 1889/1990, pp. 117–118)

Activist Nilda Rimonte (1997) describes the lasting effects of the inferiorization of the Filipino through colonization as pervasive and persistent that anyone growing up in the Philippines breathes it in with the air itself. Linda Pierce (2005) laments: "the ideologies of your family are colonized, and even your thoughts and actions are colonized, despite your initial unawareness of the systematic forces at work in the simple procedure of your daily life" (p. 32). Scholar Epifanio San Juan (1994) contends that the reality of colonial subjugation and "its profoundly enduring effects—something most people cannot even begin to fathom, let alone acknowledge" has created "the predicament and crisis of dislocation, fragmentation, uprooting…exclusion and alienation" for most Filipinos (p. 206).

The mental attitude that despises one's own and loves anything Western was reinforced by the education of the Filipino under U.S. sovereignty. Filipino historian Renato Constantino (1966) taught about reexamining our colonial history and learning from the past:

> The first and masterstroke in the plan to use education as an instrument of colonial policy was the decision to use English as the medium of instruction. English was the wedge that separated Filipinos from the past…with American textbooks. Filipinos started learning a new language but also a new way of life. This was the beginning of their education. At the same time, this was the beginning of their mis-education, for they learned no longer as Filipinos but as colonials. (p. 8)

Each of us lives our raced identities in and through the power relations that constitute our daily lives (Mac Naughton, 2001, p. 122). "White is beautiful," "White is intelligent," "White is superior," "White is powerful," "Brown is ugly," "Brown is dumb," "Brown is inferior": these were the "givens" Filipinos have been—and continue to be—immersed in. Oppressive "race" narratives such as these have become widely disseminated, taught, and repeated in our culture that they have become common sense and internalized.

Images of my growing up years often play in my mind: memories of watching aunties bleaching their skins; staying out of the sun for fear of getting dark; images of myself sleeping with a peg on my nose in my desire to have a high bridged nose; memories of hearing endless talks on people's ambitions to leave the country and live in the land of the white "masters." In settings where this kind of thinking takes place, racism has become systemic (Feagin, 2006). Certain ways of thinking and doing have become naturalized (Gramsci, 1971) to the point that they are taken for granted and therefore not open to questioning.

Migrating to Australia in 1986 was a culmination of my "white love" (Rafael, 2000) syndrome. Dreams of better life chances for my children coupled by my (mis)education led to the decision of uprooting my family from my native land. A few friends in the Philippines questioned my family's decision to migrate to Australia, a country that has been described as historically steeped with discrimination for nonwhites, warning us that settling in such a place would pose hard challenges. My response was quick: "What's so hard about it? We know that as browns we're inferior to the whites. We have no problem accepting that. All we Filipinos need to do there is work five times harder and hope that we would be acknowledged by the whites. We're lucky enough to gain approval to live in their country. Nothing like living in a white man's land."

In Australia, I strove all the more to anchor my future—and taught my children to do so—on Western ideals. I taught them to be proficient in English—at the expense of losing our home language. Most Filipinos measure a fellow Filipino's intelligence and one's possibility to gain career success by how adept a person is at speaking English. Eventually, my children lost their native tongue. With this, gradually seeped in disconnectedness to Filipino consciousness and cultural values.

Kane (2007) argues that the task of the colonist was/is to replace indigenous histories and cultures and replace them with the newly constructed racial ideologies. Culture operates as "the instrument through which the social construction of race as a system of hierarchical power relations occurs." In particular, language is the tool used to postulate racism through cultures. In Fanon's (1967) account of things, "to speak (the language of the white) is to absorb a culture" (p. 17); "it produces new ontology, a radical change in personhood" (p. 19). Kane (2007) makes these pronouncements in this respect:

> To co-opt the language of the colonizer is to co-opt racism and to "betray" one's own self and culture, and to internalise one's own inferiority. Through this historical process, the gradual loss of language and hence, culture, the history of the colonized is buried in the past: its great accomplishments and thinkers lost. (p. 357)

In the twelfth year of my stay in Australia, I enrolled in a four-year Early Childhood course. In my first three years of study, I couldn't recall any acknowledgment of different ways of thinking in any of the subjects I took apart from a Eurocentric lens. The theories that our lecturers used for explaining the image of the child, the image of our role as educators, and the image of the world were based on Western influence. Used to subordinating my thoughts and esteeming white people's ideas, I embraced the principles and frameworks that were taught in our classes and used them as the bases of my teaching philosophy in my practicum experiences. Deep inside, I felt the invisibility of my Filipino cultural experiences in my academic training. I had questions but lacked the courage to ask.

In my honors year, I was introduced to a course that changed my way of thinking. In a subject called Inclusive Curriculum, I was exposed to current debates about equity and social justice in the early childhood curriculum. Our professor and lecturers spoke about the history of colonization in Australia and traced this as the foundation of persistent racism in the country. I felt angry about the injustices and oppression brought about by British colonization. Strangely enough, at the start, I peered at this history from afar, distant from my life. Blinded by my colonial mentality that ensured that colonization would be viewed by the colonized as good and beneficial, I did not/could not see the discourse of colonization resonating with my personal history. So naïve was I of my understanding of colonialism, I used to think that having a colonial mentality was an ideal every nonwhite must aspire for.

In my honors thesis (refer appendix for further details), I examined preservice educators' perceptions on ethnic diversity and the effectiveness of their educator training in preparing them for teaching in a multicultural society. I immersed myself in literature on critical theories and antibias pedagogy. This was the start of my active journey toward unmasking the discourses of "race," racism, and colonialism that pervaded my world. I began to wake up from the dark dream of seeing myself unconsciously taking part in the violent process of colonization: raising my children through narratives of colonization. It was at this point that my interest in the "whys" of my diasporic experiences emerged. I pondered on my sense of loss, feelings of alienation, and yearnings to "come home." Although my perennial desire for my white masters' approval would often instinctively drive me back to mimic his ways and thinking, now there's a simultaneous, nagging voice that prods to resist, begs to subvert, and betray the mimicry.

The teachings of my university professors brought to the foreground the workings of "race," racism, and colonialism I have experienced in my daily life. Through them and the antiracism theories they introduced I began to understand that we are all complicitous in a system of domination and subordination, structured according to racial categories. I learned that "race" is linked to power and that racial hierarchies are normalized and made invisible.

After graduation, I taught in a kindergarten center situated in a highly multicultural suburb. At the same time, I was given the opportunity to work as a research assistant at the Centre for Equity and Innovation in Early Childhood, a research center with an international reputation for its work in equity and change, research, and professional development. My experiences in these teaching and learning communities nourished and enriched my passion for social justice education. These places created both crises and transformations in my thinking as I dialogued with early years practitioners, educators, and researchers. Over time, I have learned to retell/reread my story and redefine my identity/ties—one that speaks of agency, resistance, counter-consciousness, indigeneity, and decolonizing/anticolonialist discourse.

Reconstructing My Own Personal History: Resisting Racism/Colonialism, Transforming Racialized Identity

Where am I now in my personal journey? What I have learned since as an early childhood educator, practitioner, and researcher?

Developing a Critical Consciousness of Social Existence

First, I have learned that putting an end to oppression (including colonial subjection) requires that men and women have the ability to critically perceive their existence in the world (Freire, 2001). This starts with seeing the forces of domination and oppression that has shaped my attitudes, values, and behavior as a colonized being. It continues with naming the social and political structures that dominate and silence, and becoming aware of how the dominating structures create marginalization and inferiorization (Strobel, 2000).

In my personal journey, I have learned how institutions around me can create a structure of systemic oppression. I continue to examine and disrupt the ways in which different bodies of knowledge contribute to active colonization and racialization of nonwhite people. I strive to ensure that ongoing critical thinking form the foundation of my understandings of "race," racism, and racial subjectivities.

Taking on an Anticolonial Lens

Anticolonial thinkers have taught me to challenge discourses of social domination, interrogate works that establish dominant-subordinate connections, and contest marginalization. For anticolonial theorist George Sefa Dei (2006) anticolonial thought is about "discoursing on difference, power, racial and social oppressions as well the silences" (p. 9). For Dei, anticolonial thought is about

> a "decolonising of the mind" working with resistant knowledge and claiming the power of local subjects' intellectual agency. Resistance in this context... is about resistance to domination of the past, contamination of the present and the stealing of a people's future. (Dei, 2006, p. 11)

Building Critically Knowing Communities

In her book *Doing Foucault in Early Childhood Studies*, Glenda Mac Naughton (2005) shares a vision of educators and researchers taking active part in critically knowing early childhood communities. These maintain their "critical momentum" through linking rhizomatically, remeeting histories, deconstructing their wills to truth, reconstructing their practices for liberty (Mac Naughton, 2005, p. 211). They can be based in a locality or they can cross several localities.

In my role as practitioner researcher, I continually seek a "college of peers" (invisible and visible) who would encourage and provide me with inspiration to decenter the colonial master narratives that I have absorbed. My critically knowing communities extend to fellow Filipinos, scholars living overseas whom I knew through cyberspace, particularly when I was working on my literature review for my PhD study. Leny Strobel's (1997) work entitled "Coming Full Circle: The Process of Decolonization Among Post-1965 Filipino Americans" introduced me to the process of resisting racialization by/and decolonizing myself. The process begins with seeing myself blinded by colonial mentality and continues with understanding and overcoming the depths of alienation caused by colonization. Strobel (1997) makes this critical point:

> By transforming consciousness through the reclamation of one's Filipino cultural self and the recovery and healing of traumatic memory, the colonized can become agents of their own identity. If and when colonial mentality no longer undermines their identity, they can be an active participant capable of influencing the transformation of social structures and cultural values in the dominant culture. (p. 64)

Knowing History; Knowing Self

McLaren and Leonard (1993) speak of Paulo Freire encouraging the marginalized to question tacit assumptions—the unexamined faith in continuity and desire for familiarity—that make up the history of the oppressed, and to put under hermeneutical stress the norms these assumptions legitimate, the self-images they create, and the despair they foster. Freire's contribution has been to breathe new life into historical agency in a world that has witnessed the disappearance of the subject of history. McLaren and Da Silva (1993) explain further what this means:

> (For Freire)...historical agency is grounded in emancipatory acts as individuals challenge the everyday language and social sources to give shape and meaning to their world; it is an ongoing process involving the development of a plurality of critical literacies....In other words, the critical historical agent needs to self-consciously shape the direction of his or her desiring and the will to struggle against the decline and deformation of the possible. (p. 58)

To know my cultural history is to know my self and thus to retell a new story—a story that would be a source of empowerment, a new way of perceiving the world.

Researching My Roots: Indigenization

Decolonizing oneself involves a process of reeducating and revisiting one's culture and traditions. By immersing myself in this process, I realize that the influence of indigeneity remains in my unconscious level. In my efforts to reconstruct my ethnic identity, I learned to articulate indigenous Filipino values:

- A kapwa oriented worldview: While the English dictionary renders the word kapwa as "others," the word in Filipino implies an altogether different perception, in that it also includes the "self." Kapwa is the unity of the "self" and "others." In English the word "others" is used in opposition to the "self," and implies the recognition of the self as a separate identity. In contrast, kapwa is a recognition of shared identity. (Enriquez, 1986)
- The dialectics of kapwa as a worldview include pakikiisa (level of fusion), pakikibaka (level of fusion in a common struggle—in the face of injustice and exploitation). Pakikibaka awakens the Filipinos' consciousness of present day realities and motivates

them to be as one in their struggle to break away from the clutches of neocolonial setup existing today. (Enriquez, 1986, p. 3)

- Diwa: "Spirit"; "thread" or main thought that connects different parts; Diwa "lies at the core of our kalooban (selves), and from which emanate all personal and social sentiments. It holds together the physical and spiritual elements of existence and transforms them into one functioning whole called buhay or life." (Enriquez, 1986)
- Loob (the core of personhood): Loob has the power to shape our reality, to unite, link, and connect us to our kapwa (our fellow human being). Loob for the Filipino is a relational concept. Loob is not knowable apart from relationships (de Mesa, 1987). The Filipino "I" is not autonomous; it is "always oneself, but a self that is always inter-related.... Not the self alone, but a self in its relation(s)." (Alegre, 1993, pp. 10, 11).

Spirituality

According to hooks, "spirituality enables the oppressed people to renew their spirits to find themselves again in suffering and in resistance" (hooks, 2002, p. 117). Dei (1999) posits that knowledge, history, and politics have spiritual dimensions. Therefore the "spiritual" is crucial to antiracism praxis. Racism and colonialism bring about spirit injury—a form of spiritual, mental, psychological, and physical exhaustion that occurs as a result of battling racism and other forms of oppression where the spiritual identities of people are not integrated, acknowledged, or valued (Spencer, 2006). Faith and spirituality (though often suppressed in the academic world) are the "tools" that help me survive living within oppressive communities. The teachings of my Christian faith have implanted a strong sense of courage and hope in my life—values that have been integral to the formation of my identity and have been useful in helping me resist different forms of oppression.

Affirming Difference

> Affirming difference is crucial to anti-racism practice. To deny the power of difference is to deny envisioning alternative readings of the world. (Dei, 1999)

Difference must be celebrated and embraced. There is a need to create a space where "difference is not perceived as 'race' but as something that is valuable, something that is strengthening, and something that is absolutely necessary" (Armstrong and Ng, 2006, p. 31).

Implications for Pedagogies for/of Decolonization

I have come to learn that the works on antiracism go hand-in-hand with works on anticolonialism. What teaching approaches am I adopting using an antiracist/anticolonial prism?

I continue to critically reflect on the ways in which Eurocentric knowledge has shaped my ways of knowing and being and of my understandings of who I am and what I believe myself to be. I keep myself aware of how dominant knowledge has impacted our relationships with one another and our view of the world. The development of critical consciousness is a core philosophical tenet in this framework.

Facing history (Kailin, 2002) is another pedagogical approach that cannot be underestimated. To practise antiracism is to reclaim that knowledge that has been silenced, deformed, and hidden by those who have the power to determine whose "knowledge" reigns supreme. To face history is to rediscover our past and, as Constantino (1975) suggests, to make that past "usable." Facing history involves remeeting history—looking again at what has been and setting it in new ways that provoke us to act and see differently in the present (Mac Naughton, 2005). This entails learning to listen, engaging in multiple readings, and resisting all manners of injustice and oppression.

Finally, an antiracist/anticolonial setting must always be a "decolonizing space," a space for "social, spiritual and intellectual enrichment" (Dei, 2006, pp. 311, 313), a place where inclusivity is fostered, issues of identity are addressed, transformative learning is sustained, and hegemonic paradigms are challenged and disrupted.

CHAPTER 8

Adults Constructing the Young Child, "Race," and Racism

Sue Atkinson

Between 2003 and 2007 as an Indigenous Victorian working in the early childhood field, I interviewed forty members of the Indigenous and non-Indigenous early childhood community in Victoria to gather data for my PhD thesis "Indigenous Self-determination and Early Childhood Education and Care in Victoria." Research participants were interviewed around five broad questions. It is the responses of the research participants around one of those questions "What Understandings of Victoria's Indigenous Cultures do Indigenous and non-Indigenous Preschoolers Hold?" that is the basis for this chapter. In analyzing participants responses, I concluded that Indigenous and non-Indigenous children are exposed to colonial concepts of the Indigenous "other" imbedded in dualisms such as Indigenous/non-Indigenous, assimilated/traditional and that these understandings inform the relationships between Indigenous and non-Indigenous children and the emerging identities of Indigenous children in early childhood settings.

Intersecting with these childhood discourses of Aboriginality I found the adults' discourse of the child as innocent of "race." It is this discourse that acts as a barrier in addressing "othering" within early childhood programs.

Adult Discourses: The Young Child as Innocent of "Race"

Within the construction of the young child as innocent of "race," the preschool child unlike older children and adults is unable to use abstractions such as "race" in a meaningful way, such as making decisions regarding who they will interact with in early childhood settings. It is thought that young children may use racial terms in an imitative fashion but with little understanding about or reflection on their meaning on a social level (van Ausdale and Feagin, 2001).

The construction of the preschool child by Indigenous and non-Indigenous adult participants in this study echoes this understanding. In talking about the young child in terms of racism/prejudice or bias, the majority of participants' responses were couched in terms of absence and innocence. Such responses can be seen through the lens of the child as color-blind in simply treating other children as human beings without the adult trappings of "race" and racism.

As Marilyn a Koorie Early Childhood Field Officer (KECFO) states, "A four year old is a baby and they aren't going to see any difference." Children were positioned as ignorant of skin color including their own, until it was identified by adults, as Celine an Indigenous early childhood professional observes:

> Usually children don't know what color they are until someone points it out and says you're a black fella or white fella then when these things are pointed out they realize and start asking questions.

Although examples of bias against Indigenous children in non-Indigenous spaces were reported, they too were positioned within a context of innocence. As Maryla a non-Indigenous kindergarten teacher relates:

> That sometimes happens where a child will say "I'm not playing with them they're black" "I'm not playing with the black child" and I've often found that if we investigate, it will be more often about what is happening within that set of circumstances, that there might be a dispute about a piece of equipment or an object or an area to play in that we then address rather than it really being an issue about the color of someone's skin.

In this situation, there was a denial that the white child could be using racial awareness to negotiate his social space by excluding the black child. David an Indigenous parent whose niece was the object of racially based exclusion agreed that the experience was informed by childhood naivety:

> One of the kids at the pre-school didn't want to hold her hand because she was black...It was not racism it was (a reaction to) difference through naivety and curiosity

Faith a non-Indigenous CSRDO (Children's Services Resource Development Officer) who supports Indigenous inclusion in child care centers by providing resources explains white children's negative reactions to these resources in terms of ignorance:

> I have known of biases with children who didn't want to play with black dolls, feeling awkward about touching a black doll. When a center says "We don't have any children of different backgrounds here just the white Anglo," when you take different resources, the black doll or black posters and often children aren't sure of what to do and that may not be racism just that lack of awareness that children think everyone's white and not being exposed to diversity.

In spite of their assumed innocence and ignorance these preschool children are recognizing difference, making judgments, and acting on these in rejecting the black doll and the black child. Childhood curiosity in the above examples doesn't extend to interacting with the black child or exploring the black doll.

Adult Discourses: The Child as Imitator

When it was acknowledged by participants in this study that children did use racism in their interactions with other children the influence of adults or older children was employed to explain the child's "aberrant behavior," with children merely mimicking what they heard at home. As Llewellyn, an Indigenous parent and community leader states:

> I know it [racism] happens...I haven't seen it with Sue or Natasha but in my job I know it happens I also know with my older daughter, she's twenty-one now. When she was in preschool I know it happened, it happened to her. But all kids do is parrot what other people and adults say

But by applying what adults say regarding colonial understandings of Aboriginality to specific situations in acting out their whiteness on black children suggests that children are active in understanding beyond mere mimicry (van Ausdale and Feagin, 2001) and that this understanding is moving from adult to child.

Tensions in Adult Discourses of the Child: Fracturing Discourses of Innocence and Imitation

Young children were often positioned by the participants in this study as innocent in not seeing, understanding, or knowingly employing racial hierarchies. But non-Indigenous children were also described as understanding Aboriginality as undesirable and employing these understandings in a social context to marginalize Indigenous children. Although participants sought to foreground both the Indigenous and the non-Indigenous child as innocent, there is a sense of fracture around this construction.

Participants such as Anita (a KECFO) and David (a parent) who positioned children as largely innocent of "race" hold tensions within their own understandings. Anita positions racial understandings on an "unconscious" level, where the connections between observations of difference and the social meanings of "race" are yet to emerge:

> They [preschoolers] are beginning to see differences around them and at grade two or three you realize that difference is Aboriginal at a real conscious level.

Yet when discussing the benefit of Indigenous inclusion in early childhood programs Anita noted:

> The chance to change non-Indigenous children's views at that early age about Indigenous people I'm always conscious that when I do go out to a center that I am probably the first Aboriginal person the children may have met and this experience will be positive and they will always remember that.

Here Anita is suggesting that although non-Indigenous preschoolers are unlikely to have met an Indigenous person, they will probably have formed an opinion of Aboriginality and it is probably negative.

Although David positioned the concept of children and racial bias at a preschool level as being complicated, he concluded that children at this age were largely unreflective in imitating adults around them. In spite of positioning racial bias as largely imitative David saw these concepts as firmly established by the first years of school and actively informed by colonial constructs of Aboriginality:

> When kids get into primary school its more that kids have got set ideas, kids in grade prep and grade one think they know everything and if you don't fit in that box you can't be Aboriginal "You don't have curly black hair, you're not black so you can't be Aboriginal."

In their constructions of children and "race" these participants are exercising a reflection that can be described as postcolonial in its ambivalence. There is an impulse toward recognition of children's understanding of "race" at one moment and denial at the next. The movement back and forth between these two positions is a postcolonial understanding, fluid and entangled with contradictions.

Although colonial understandings of the child as innocent largely dominated the participants' understandings around children and "race" in the preschool years, it was recognized that colonial concepts of Aboriginality were emerging in these early years to become evident in primary school.

Angel (a parent) wonders if the differential nature of the preschool and school environments in terms of supervision may allow children's biases to come to the foreground in the school play ground:

> In the class room, her and her friend Tammy, they are both dark, one of the kids called them "coon" or "boong" or something. The school dealt with it really well. It happened in the play ground, but maybe within the preschool...it's a close intimate environment and you have a teacher and assistant there and there are only twenty kids in the class...it's [school's] not as heavily supervised in the playground... there are only one or two teachers on yard duty.

When children hold racially informed concepts of others during the preschool years and are aware that the open expression of these concepts will met with the disapproval of staff they may be less likely to express them when closely supervised. The workings of racialized power between children may therefore play out on the periphery of adult awareness (Skattebol, 2005).

Recent studies that have engaged with such tensions around young children and "race" as raised by participants have consistently challenged the concept of the child as innocent. Mac Naughton (2005) points out that young Australian children recognize skin color and use this recognition in making decisions around self worth and the worth of others.

In exploring non-Indigenous children's views about Indigenous people through the Pre-school Equity and Social Diversity Project (PESD) Mac Naughton (2005) found that Anglo-Saxon preschool children employed colonial understandings of Aboriginal people as "the other." She writes:

> No child in the PESD group shared any information that suggested that Aboriginal-Australians and Anglo-Australians have anything in common...instead, they had learnt that Aboriginal people were odd and different and at times bad. Some children had also learnt they were to be feared. (p. 170)

Non-Indigenous children's framing of Indigenous children as the distant exotic "other" is expressed by Franka, a kindergarten teacher even as she overlooks this perception:

> If we are focusing on the bush and Koorie culture and things like that the children ask a lot of questions about how did they use to kill animals then you talk about boomerangs and spears and where did they live and where did they get their other food they will often ask did they have Macdonald's but it's hard to say what their perception is.

Not only do white children position Indigenous people as the exotic, strange, and at times fearful "other," the objectification of the black "other" with its dualistic desire for white is also expressed by Indigenous children. As Mac Naughton (2000) in outlining fifty years of research on children and "race" writes:

> Black children consistently choose white dolls as those they'd like to play with and as the ones that look pretty (Gopaul-McNicol, 1995). Black children showed a bias towards lighter skins (Averhart and Bigler, 1997) and black children showed racial preference to whites. (Kelly, Duckitt, and Johnson, 1995, p. 9)

Some of the participant's responses also demonstrated the tensions within the ideology of innocence/ignorance with Indigenous children employing colonial constructions of Aboriginality as undesirable or frightening. As Lee, a non-Indigenous teacher in an Indigenous space notes:

> They are not aware of it yet, they are not aware of racism, yet. It's really funny though really dark-skinned children are afraid of the Koorie dancers because they are black and they make comments about it "Oh they are really dark!" and they cry and carry on.

Two Indigenous parents also described conversations where the desire for white was raised by Indigenous children. Celine reported a conversation she had with her three-year-old son whose skin is fairer than her own:

> My son says "You've got to get brown like me." I say "I am brown, I'm black" "No your black mum you've got to get brown like me" and he's only three years old.

Angel similarly reports:

> I know another non-Aboriginal mother who has an Aboriginal child in preschool and the daughter says to her "I don't want to be black I want to be like you mummy" and maybe it's a mother daughter thing, where daughters just want to be like their mothers.

Not only are the racial attitudes born of colonization excised by white children in their relations with Indigenous or black children, but Indigenous children operate within its dynamics to judge themselves and their families in racist ways where white is desired and blackness derided.

Colonial Understandings, Skin Color, and the Complexity of Aboriginal Identities

Although the children who experienced marginalization as described by participants were dark-skinned Indigenous children, fair-skinned Indigenous children also had to deal with colonial understandings of Aboriginality in early childhood centers. David the father of fair-skinned Indigenous children described the challenges they faced.

> If one of the kids say they're Aboriginal and they [non-Indigenous children] say "You can't be you're not black" and that sort of thing. Or they might go home and say something to their parents and come back and say "Mum says you're not."

These challenges arise from the confrontation of non-Indigenous people, (children and parents in this case) with a postcolonial Aboriginality that is discordant with colonial constructions of Aboriginality as black.

The colonial construction of Aboriginality as black is employed here to deny fair-skinned Indigenous children their Aboriginal identity. While the white child exercises his/her power in defining who is white and who is not based on skin color, the fair-skinned Indigenous child begins to grapple with the complexities of that identity as he/she formally journeys into the mainstream via the preschool.

Primose, an Indigenous parent and early childhood educator describes her daughter's confused response to the preschool curriculum:

> She got a bit confused because her first introduction at preschool was of dark people painted up... so when she was coming home she was speaking about the Aborigines in the central desert and how they looked for bush tucker and that sort of stuff. It wasn't until Anne went in. And when she saw Anne it clicked to her that she was Koorie like Anne who has fair skin too. It was a shame actually.

Although a more sophisticated understanding of Aboriginality has developed with the partnerships formed between local Indigenous communities and non-Indigenous early childhood centers in Victoria, as Fabian a KECFO states:

> I think teachers are getting more understanding and more awareness of Indigenous culture and history where before they didn't. The children and the parents, with the KECFOS going out to those centers, they are learning from us as well and I think that's for the betterment of the teachers and children as well.

The representation of authentic Aboriginality as dark skinned and distant persists. Even as "racist incidents" are dealt with in the early childhood setting, they may work to reinforce colonial understandings of Indigenous identity. Garry an Indigenous parent describes his involvement in such an approach:

> They talked about Indigenous Australians one day and I ended up going out there and taking some stuff from home and taking the didj [didjeridu] and playing it.

Garry noted that the children already had knowledge of Aboriginality based on colonial representations:

> Even the kids and staff were, you know asking questions about bush tucker and all those sorts of things. It's like trying to explain there is a whole contemporary Aboriginal society it's not just caught back in history!

Non-Indigenous educators' understandings of Aboriginality are largely positioned around the binaries of black and white (Davis, 2005). Indigenous early childhood centers provide an alternate framework for identity where both fair- and dark-skinned Aboriginality are normative. As Aunty Roslyn describes:

> The difference of the skin, the thing about it, it doesn't really matter whether you are really dark...[or] blonde haired blue eyed...It doesn't really matter. We are all the same...Koorie children will let you know they are Koorie no matter what color the skin. It just seems to be that you teach them you are still accepted whatever color you are.

Yet fair-skinned Indigenous children are not always spared challenges to their identity in Indigenous early childhood spaces. Heather described her fair-skinned Indigenous child's confusion and distress when targeted "as a white racist" by another Indigenous child:

> There was one kid who was really dark whose mother was having a really hard time at work with racism...So this beautiful kid was picking up on all this and drawing her own conclusions and was really anti-white...Sue got caught in the cross roads. Even though this kid knew about Sue's own family background but at the same time as far as she was concerned Sue was still white...That's what triggered off Sue's little dilemma in terms of identity...Sue was repeating back all the things that this other kid had told her that she was hearing at home and at the same time Sue was internalizing some of that as well.

Not only is the angry dark-skinned child targeting Sue as white, even though she "knows Sue is Indigenous," she is conflating white with oppression. One of the greatest insults an Indigenous person can make to another is to call them "white" (Dudgeon and Oxenham, 1990). In the struggle to reclaim and recover an identity that has been attacked and eroded under colonization, colonial stereotyping is inverted, with black as beautiful and white as morally and culturally inferior (Loomba 1998). Learning what it means to be Indigenous for Indigenous children is paired with learning that being white means having the power to oppress and to marginalize.

The marginalization of Indigenous children as described in these early childhood spaces reflects the experience of the "outside" world for Indigenous children and their families (van Ausdale and Feagin 2001). Early childhood centers are not partitioned off from a racist society and are often placed as replicating the racism of that society.

The Replication of Systemic Colonial Discourses of Aboriginality and Early Childhood

That constructions of Aboriginality continue to be informed by colonial concepts of racial hierocracies and is clearly expressed by Anita a KECFO: "If you look at the hierarchy of races in Australia, Aboriginal people are on the bottom!"

Although such discourses that position Aboriginal people at the bottom of a hierocracy (Moreton-Robinson, 2000) and around the binaries of white/black, civilized/uncivilized, desirable/undesirable, superior/inferior (Davis, 2005) are tenacious, they are often also insidious, as Aunty Roslyn describes:

> I reckon there is always racism. It's underlying and it's very hard to actually pick, it's there and you can't really say anything about it because it's not right in your face, it's underlying.

Mac Naughton (2000) argues that young children's construction of "race" is linked to these colonial discourses in diffuse and intermittent ways, creating discourses of "race" that demonstrate the interconnections between the social worlds of child and adult. Van Ausdale and Feagin (2001) also argue that children are exposed to concepts of "race" and racism from the moment they enter society. This ubiquitous yet often invisible racism impinges on children's understandings of "race" well beyond the immediacy of the family.

In underestimating the ubiquitous nature of racism within Australian society it may be viewed as an isolated problem due to the acts of a few errant individuals who act as role models for their children. Early childhood professionals need to be more aware of the complexities around understandings of "race." Burnett (2001) describes postcolonial racism as "fluid and evolving," and calls on early childhood educators to recognize "the imprecise and blurred make-up of racism if it to be addressed more effectively" (p. 105). Dealing with children's constructions of "race" that fail to take into account the pervasive nature of racism can perpetuate colonial understandings of "race" that unintentionally support individual children's biased actions. In excluding, marginalizing, or misrepresenting Indigenous culture within the early childhood program, early childhood educators fail to take into account the influence of the early childhood center itself in perpetuating colonial understandings of "race."

Indigenous Responses to Issues of Children's Identity and "Race" within Early Childhood

Although the continuing colonization of Indigenous children is silenced with the denial that they are subjected to "real" racism in the preschool, it is recognized that Indigenous children need to prepare for a life of dealing with racism that nominally begins at school (Walker, 1993). Garry describes what he thinks his daughter will have to face:

> The school will be pretty good and the teachers will be good but it's the other kids who either have had plenty of interaction with other cultures or haven't. You might have this issue of kids who might be racist and kids who aren't I'm sure she's going to bump into plenty of kids who . . . [pause] . . . she's got quite brown skin.

Indigenous inclusive programs in early childhood that enhanced children's identity were positioned as central in assisting Indigenous children to deal with the racism they would inevitably encounter. Indigenous-specific early childhood spaces were positioned as particularly relevant in enhancing Indigenous children's identity in ways that challenged colonial constructions of Aboriginality, as Debbie an Indigenous early childhood professional relates:

> It strengthens their identity and gives them confidence because they are told at such an early age "You are Koorie you're beautiful" all those positive words.

Indigenous children may struggle to develop a healthy sense of self within a racist society. Indigenous children may be covertly learning about their position as the racial inferior within non-Indigenous early childhood spaces, while their families are trying to engender Indigenous pride and construct a state of mind that prepares them to deal with racism (Joseph, Lane, and Sharma, 1994). As racism in the early years is overlooked or dismissed in the context of innocence so may be its impact on its young victims.

Implications for Antiracist Pedagogies: Challenges for the Field

In light of the findings of this project, there are a number of ways that early childhood educators can work to confront and challenge the racism faced by many Indigenous children and families and the discriminatory beliefs and practices of many non-Indigenous adults and children. This antiracist pedagogy, while developed out of the work with a local community, has implications and possibilities within an international arena, especially in nation-states with Indigenous populations.

Growing Strong Identities: Children as Activists

It is hoped by the Indigenous participants in this project that the sense of Indigenous pride engendered in the home and in the Indigenous early childhood center will empower Indigenous children to deal with racism. As Anne a parent and early childhood professional states:

> If they are brought up being proud of who they are when they become preschoolers hopefully they will have the courage to say "No, that's not Ok I don't like what you are saying."

Teaching young children social skills such as "using your words" is positioned as a priority in early childhood. Encouraging children to express their feelings when other children are acting out of bias can build assertiveness on the part of the "victim" and empathy on the part of the "perpetrator" (Derman-Sparks, 1989). But these skills are often taught within the framework of innocence, in that, children are positioned as unskilled in reflecting on their own actions and largely unaware of the implications of their behavior on their "victims" (van Ausdale and Feagin, 2001).

But van Ausdale and Feagin in their 2001 American study demonstrated that young children use racial language to actively exclude other children based on "race." In an exercise of power, racial language is used by white children to control how children of color use space and materials in the early childhood center. In Australia, Taylor (2005) also demonstrates how Anglo-Saxon children can use their awareness of difference to exercise power over racially different children to exclude them from play.

Taylor sees the need to challenge the discourse of innocence in early childhood. Denying that children exercise racialized power in a conscious way will not undermine the colonial discourses of "race" reflected in such interactions. If early childhood educators in Australia continue to invest in the concept of childhood innocence, the provision of programs that deal with "race" and racism such as antibias programs are less compelling. In reconstructing our understandings of young Australian children and "race," we can then begin to build a postcolonial knowledge of Aboriginality with young children with conviction.

Such knowledge can build empathy and activism in non-Indigenous Australian children and is foundational in really hearing Indigenous children's words as they negotiate racially charged social interactions in early childhood spaces.

Deconstructing the Discourses of "Race" and Identity in the Early Years

Two forms of colonial discourse dominate this chapter: adult's discourses of children's innocence around "race" and children's discourse of Aboriginality. Although adult discourses of children and "race" are largely constructed around binary oppositions such as child/adult, innocent/knowing, imitator/agentic, this chapter also reveals tensions and contradictions within these binaries. Even though fractures in the discourse of childhood innocence may be muted when participants were responding to direct questions about children and "race," they are strongly articulated in Indigenous participants' descriptions of the colonial concepts of Aboriginality that Indigenous children confront in the early years of education and care.

Such concepts of Aboriginality rest on binaries where Indigenous is the opposite of, and inferior to, the non-Indigenous, or more specifically the Anglo-Celtic. This produces "others" who are different from and inferior to the Anglo-Celts, who dominate Australian society.

Mac Naughton and Davis (2001) recommend challenging otherings that simplify complex racial identities and that overlook the intimate connections between Anglo-Saxon and Indigenous in addressing racist ideologies with young children. It is acknowledged by the participants in this study that adults are a primary source of information about "race" for young children. Mac Naughton and Davis (2001) recommend that early childhood educators employ theories of deconstruction such as Derrida's (1973) to provide opportunities to challenge such binary oppositions as Indigenous/non-Indigenous and child/adult by exploring the excluded middle that positions identity and knowledge in more fluid and complex ways. In the exploration of the space between the racial awareness of child and adult, and the stereotyped and lived realities of Indigenous families, Derrida invites early childhood educators to explore what has been silenced or hidden and to focus on the complexities that children and "race" present. In employing Derrida's theory, an Indigenous child can be white, a young child can be racially aware and a genuine victim of racist exclusion and its intersections in the early childhood center.

Reconstructing a Discourse of Racial Identities and the Child

Mac Naughton and Davis (2001) identified that children who were in programs that actively built postcolonial knowledge around 1) the multiplicity of Indigenous identity and 2) the impact of colonization on Indigenous people demonstrated nascent understandings of Aboriginality. These understandings reflected the current political and social position of Indigenous Australians. Such understandings can be built on subjugated voices from the "excluded middle" (Derrida, 1981), where Victorian Indigenous children and families live, challenging the dichotomies around Aboriginality and building more complex understandings of racialized identities.

The early childhood center can become a site for postcolonial narratives of Aboriginality in opening up a space for Indigenous voices and for dialogues in which Indigenous people define their own identities in ways that defy the black/white, authentic/inauthentic dichotomies by naming what is hidden and disrupting these binaries. It is through constructing new knowledge with the Indigenous community that early childhood professionals can reveal the connections between Indigenous and non-Indigenous children as they navigate their racialized identities and provide the foundations to disrupt the reproduction of racism.

Beyond the Discourse of Childhood Innocence and "Race"

In positioning the young child as largely innocent of "race," both the dark-skinned and the fair-skinned Indigenous child continues to deal with colonial constructions of Aboriginality in early childhood spaces largely on their own. In denying that colonial understandings of Aboriginality find "real" expression in the social relations of early childhood, we allow colonial constructions of Aboriginality to bloom in early childhood. The fair-skinned Indigenous child is excluded as Indigenous, the young child is unacknowledged as employing concepts of racism, and the dark-skinned Indigenous child is excluded from active adult support in dealing with the racism that excludes him/her from full participation in the early childhood curriculum.

Although there are contradictions between adults' assessments of children's understandings of "race" and examples of children's racially based knowledge and actions within this study, it is at least recognized that children's concepts around "race" are developing within the preschool years and are formed by the early years of primary school. The need to address the silences around "race" and racism in the early years is therefore essential, regardless of children's perceived innocence, as it is the early years that are foundational in building the platform for social justice to which early childhood educators are ideally committed.

CHAPTER 9

Languages Matter: My Subjective Postcolonial Struggle

Prasanna Srinivasan

> Furthermore, the colonized's mother tongue, that which is sustained by his feelings, emotions and dreams, that in which his tenderness and wonder are expressed, thus that which holds the greatest emotional impact, is precisely one which is least valued. It has no stature in the country or in the concert of peoples. (Memmi, 1967, p. 107)

My Subjective Dimensions—Disruptive Beginnings

I am an Indian, who migrated to Australia a few years ago with a young family. I grew up in India, during the postindependence era, free from the clutches of colonialism with the fire of being a "resistant colonized subject" still fuelled and kept alive with ideologies of anticolonialism. My history books talked about my "Indianness" as being complex—historically, culturally, linguistically, religiously, and economically and the ways in which the colonizer (*vellaikaran*—white man in Tamizh) repeatedly tried to unify us in the name of civilizing the natives, especially with a Western language and a Western system of education. As a resistant colonized subject, I considered that being affiliated to any colonial identity would mean succumbing to the colonizer's dominance. I negotiated my interactions with Australian colonial whiteness with deliberate consciousness of my subjectivity, as I carried the guilt of being the "resistant subject," with a predisposition to suspiciously challenge or depose anything identified as colonial. Therefore, I resorted to silencing myself to suppress any feelings of retaliation. When I was asked whether I spoke Indian at home with my children or what curry I cooked for dinner, I angrily thought there is no language that is Indian, nor is there an Indian dish called curry. I blamed such thoughts on my resistant subjectivity surfacing to question *vellaikaran*'s innocent curiosity and began to silence myself guiltily.

My children attended early childhood services and soon were engaging with one of the rudimentary socioeducational settings in Australia, which to me seemed to be espousing the culture of the colonizer. However, it was just normal, to the Anglo-white early childhood educators who interacted with my children, with their aim to effectively socialize our children, to be "normal." They said that my children's "language skills" were developing satisfactorily and I should be very happy that they have "learnt to communicate." They reassured that the continual use of "Indian" at home will ensure the maintenance of my language in them. I disappointingly thought that my children *had* language and communication skills and even though their skills in English had picked up, their interactions in Tamizh were diminishing. I worried that my children would eventually become steeped in colonial expressions and become far removed from our ethno-linguistic identities. I was not willing to allow our expressions to be superimposed by those of the colonizers. We vehemently persisted in instilling our language in our children and I suspected such persistence was my resistant subjectivity suspecting *vellaikaran*'s efficient tutelage, and my guilt-ridden silence continued.

My Subjective Determination—In Search of a Multilinguistic Utopia

I began studying and working as an early childhood educator in Australia (Melbourne) and I actively tried to engage in practices that encouraged the development of children's first language. I spoke in their first language if it was one of the languages I spoke (Tamizh, Hindi), but after a few weeks in the service, the children would avoid speaking to me if I spoke to them in those languages. However, they would readily engage in discussions with me, if they were in English. The theories of early childhood insist these years to be a vital period for children's language and identity development (Fleer and Raban, 2005; Siraj-Blatchford and Clarke, 2000) and that children's ethno-linguistic backgrounds should be strongly supported during early childhood. It is also emphasized that children become better skilled in learning English only if the development of their primary language is fostered (Corson, 1993; Cummins, 1989 and Ramirez et al., 1991, both cited in Nai-Lin Chang and Pulido, 1994; Skutnabb-Kangas and Toukamaa, 1976 cited in Makin, Campbell, and Diaz, 1995). English being the medium of instruction in most Australian schools, it is considered that failure to pick up reading and writing skills when young in English has a negative impact on their success in schools (The Parliament of the Commonwealth of Australia, 1993; Hill, 1995 cited in Raban and Ure, 2000). Informed by these theories, I thought that it is less problematic when the children's linguistic background is the same as the dominant one, English, as the current environment is already structured to provide that support. On the other hand, this demands bilingualism in learners of English as Second language (ESL). Therefore, matching the linguistic backgrounds of staff and children seems to be the only way to support and develop first language so that children develop proficiency in their own language before being exposed to the dominant language, English. Developing bilingualism is even more critical for the Indigenous people of Australia, as Baumann (cited in Joliffe, 2004) states, that language is the core identity of Indigenous people and bilingual education has been enormously successful for their community. Baumann adds that children's cognition is linked to their first language and they are able to transfer outside knowledge using this initial learning. Pearson (2005 cited in Van Tiggelen, 2005) adds that many languages thrived in precolonial Australia, because most Indigenous groups were multilingual. I sought support from agencies to access bilingual staff, to enable children to continue using their first languages, whilst learning ESL in early childhood services. However, I learnt that funded support was provided only if children were seen as having difficulties in communicating with staff and peers but not specifically to maintain their first languages. I was also given handouts about those ethnolinguistic cultures to widen my understanding of those cultures, including mine. I lamented at the purposes and inadequacies of funding and notions of representing such cultures by writing about them, as I couldn't see the complex me being represented by these handouts about India. I felt compelled to investigate this further by engaging in a research study (refer appendix for further details) in order to interpret current understandings of including cultures and languages. I contemplated whether such compulsion was due to my resistant subjectivity denying *vellaikaran*'s magnanimous contributions to aid cultural understanding. Nevertheless, my guilt-ridden silence took a turn, as now I researched to express the colonizers' understandings of languages.

My Subjective Collisions—Histories, Theories, Policies, and Discourses

My research study involved gathering early childhood educators' understandings of including linguistic diversity in early childhood settings. I was now engaging deeper with literature surrounding the topic of linguistic diversity and I discovered the term "linguicism." Linguicism is, like racism or sexism, a bias that needs to be challenged, as it predisposes the native speakers of the dominant language automatically with power and superiority, creating unequal distribution of resources and feelings of inferiority among those who have to learn English to function within a society (Arthur, 2001; Grin, 2005; Skutnabb-Kangas, 1988). Education to teach only English is seen as a form of bias called "institutionalized linguicism" by Skutnabb-Kangas (1988) as it trivializes speakers of other languages. Political discourses such as the endorsement of a national language and the introduction of citizenship test are then a form of bias. According to Calvet (1974 cited in Young, 2001), the dominance and hegemony of colonial power, its language and culture is reinforced and appropriated by the governing body through its political discourse. As Foucault (1998 cited in Mac Naughton, 2005) states, it is important to examine past events and histories to understand the role of the past in molding the present. Historically, as noted earlier, Australia had been a land of many languages before colonial occupation (Clyne, 2005; Van Tiggelen, 2005). However, since the days of colonial settlement, the political language of Australia has been English, and therefore it has become the language of powerful discourses. The validation of citizenship test (Commonwealth of Australia, 2006) includes English proficiency test and appropriates dominant, Western, monolingual discourse. Therefore, the discourse of political propaganda of nationalism through legislations that uphold the dominance of English in a country like Australia, whose linguistic multiplicity is widening, can only result in rendering some more powerful due to their inheritance of that linguistic capital. I progressed to examining two key political documents, "The Multicultural Agenda" (Department of the Prime Minister and Cabinet, 1999) a document drawn by the Australian government to promote its commitment in promoting cultural diversity and enhancing harmony, productivity, and equity for all Australians. "The Quality Improvement and

Accreditation System" (QIAS) booklet, published by The National Childcare Accreditation Council (NCAC, 2001) on the other hand, outlines principles that define quality of care in early childhood settings. Some QIAS (NCAC, 2001) principles address cultural and linguistic diversity and outline guidelines on supporting children and families from diverse ethno-linguistic backgrounds.

I now summarize the key ways by which the policies I reviewed suggested how the Australian political body seeks to maintain colonial control through the colonial language, English, over the rest of the linguistic groups in Australia:

- Asserts English, a colonial language, as the national language of Australia. Thus overrides Australian indigenous identity and the linguistic identities of many to secure national identity.
- Establishes the superiority of English, yet promises to treat all alike with respect and equality.
- Suggests respect for all ethno-linguistic groups on one hand, but aims to propagate English in the name of participation and success on the other.
- Aims to provide and maintain support for some languages, in some spaces, for the prosperity of all in the nation and not really for the intrinsic values of all languages.

I wanted a voice or a language to express the discourses of including languages and cultures in Australia, as until now I had silenced myself from overtly questioning these espoused notions. I related to my past subjectivity, a resistant colonized subject, and could not verbalize my objections due to feelings of guilt and fear. I continued to study and work in the field of early childhood; my discomfort with the language used to address ethno-linguistic differences never ceased to abate. I felt compelled to express and needed a theoretical base through which I could articulate my personal inquiry with concrete fundamental concepts.

My Subjective Cohesions—Postcolonial Language

During my masters study, I was introduced to postcolonial theories by my mentors, as I struggled to find coherence between my silenced guilt juxtaposed with my desire to express. Postcolonialism can be defined as the period that comes after the domination of colonial dominance, as well as one that examines the residual, economically driven colonial dominance still in circulation (Young, 2001). Young adds that postcolonialism also aims to act against the tacit domination of colonial power in socioeconomic and sociocultural aspects of the society. Thus, as I had always resisted colonial dominance, I was drawn toward postcolonial theoretical framework due to my desire to challenge the colonizer's imposition through English, the colonial language. Moreover, it also enabled me to conceptualize vociferously my silenced discomforts about the language of ethnolinguistic inclusion. The use of postcolonial theories of "Orientalism" (Said, 1978) and "critical white theory" (Frankenberg, 1999) became pivotal to my study, with which I was able to question the relevance of grouping, and simultaneously justify the need for grouping to recognize dominance. Postcolonial framework enabled me to deconstruct two major terms that are still being used to address languages and ethno-linguistic groups: Languages Other Than English (LOTE) and Culturally and Linguistically Diverse (CALD).

Postcolonial Language—Questioning the "Silent Dominant"

The following key postcolonial concepts enabled me to verbalize my silenced discomforts, which I believed were to do with my resistant subjectivity. These concepts revealed why I was troubled when asked by Anglo-white Australians whether we spoke "Indian" and sanctioned us to speak "Indian" in some spaces, whilst teaching my children to be "normal," and most of all grouped the rest of the ethno-linguistic groups and remained unnamed.

Othering and Languages in Australia

This postcolonial concept describes a pivotal concept repeatedly used by the colonizer (self) when engaging with those that are different (other). It is described as the discursive practice of using texts about the colonized by the colonizer to authenticate their power and authority (Mills, 2004), and these texts are used not just by individuals but also by institutions. According to Said (1978 cited in Mills, 2004) the "other" is created and spoken for negatively by the colonizer to position self as being more superior and sophisticated. Young, (1995 cited in Mills, 2004) suggests that Indigenous peoples in colonized lands were linguistically categorized and grouped forcefully, resulting in unequal distribution of wealth. Mills, (2004) adds to this fact by saying that such grouping of varied people makes their identities dispensable and less worthy and the "other" is addressed and spoken for from outside by the colonizer. According to Foucault, (1981 cited in Mills, 2004) linguistic descriptions are used to label items of the world, although they are not naturally ordered. Foucault adds that categorization and naming creates distinct groups with stringent boundaries, although they may share commonalities, and the dominant system erodes the knowledge of the rest.

In the case of Australian languages, the sociopolitical categorization of languages through the establishment of English as the center and the rest, including Indigenous languages, as LOTE has marked distinct grouping and domination of one over all others. It was noted through the examination of policies that the Australian government speaks for funding "other" group to learn English as if this group is deficient. Viruru (2001) and Gandhi (1998) discuss how the colonizer asserts and reproduces his/her dominance through the imposition and propagation of his/her language. As discussed earlier, Australian government has stabilized "the colonial self" and legitimized its propagation by making English as the national language. Moreover, it allocates funds to promote some languages in certain spaces and this can be seen as an act of maintaining "other" to empower "Other" (self).

Normalizing and Languages in Australia

This is another colonial discourse that enables the colonizers to impose and propagate their identity in those that are different. The colonizers usually engage in the discourses of "othering" and "normalizing" simultaneously. Whilst othering is used to think and talk about those that are different, normalizing is used to think and talk about self. Critical white theory (Frankenberg, 1999) discusses the evidences of colonial discourse that aims to mask the dominance of self, thereby creating an invisible and neutral center. In her discussions about recognition of dominant "race," Frankenberg (1999) highlights the importance of questioning centralizing and normalizing "White" by attributing invisibility to this group. Thus Frankenberg (1999) validates respectful recognition of differences that aim to challenge dominance and injustice in order to create equity. Although this theory recognizes the assertion of superiority through differentiation and categorization as an essentialist colonial discourse, it also establishes nonrecognition of differences as a discourse that reflects covert colonialism. Therefore, as a first step toward avoiding othering, Frankenberg (1999) authenticates the recognition of "White culture" by self.

Culturally and Linguistically Diverse, "Community languages" and "Home languages" naturalize English and categorizes the rest as special or strange. These are the terms used widely in Australia, especially by the wider sociopolitical institutions, to box many cultural and linguistic groups present in our society. However, Anglo-Saxon as a cultural group and English as its language is not only excluded from this group but is also unnamed otherwise, thereby making it normal and invisible. This group has a culture and language, and through its linguistic discourses, circulates and propagates its value system without being questioned in Australian society. One should also consider English as one of the languages that is practiced at home and in the community, and above all, English as a settler or migrant language, which was introduced to Australia since colonization. The nonrecognition of English as a language predisposes unquestioned support and enables the funding allocated to institutions in the propagation of one set of values to be overlooked. Such invisibility has resulted in hiding the magnanimity of their ethno-linguistic practices dictating our lives. Moreover, it has resulted in reducing the ethnolinguistic practices of the peripheral groups to be simple—translations and cue words that some fund to enable their sustainability.

Thus, postcolonial language enabled me to recognize and question colonial discourses of othering and normalizing. However, this did not end my subjective strife, as I felt urged now to depose all that I recognized as being affiliated to colonial discourse.

My Subjective Shifts—from Resistance to Resilience

My engagement with postcolonial theoretical framework enabled me to identify the colonial discourses that were still in circulation in the sociopolitical arena of Australia. Nevertheless, this also resulted in two opposing dispositions within me. On one hand, it empowered me by providing me with a specific language to challenge colonial ideologies about languages and cultures. On the other hand, this also made me want to refrain from using discourses that I now identified as being colonial. First, as a resistant colonized subject I wanted to refrain from using the language of the colonizer. I felt at a loss after having found a language to challenge the colonizer. Second, I didn't want to use the terms LOTE and CALD that personify the colonial discourse of othering and normalizing. I primarily did not want to see myself and be seen as the "other" or "diverse," as I recognized the meanings attached through a postcolonial lens. Thus, the use of postcolonial platform did not end my subjective strife.

From the very beginning I had always questioned the place of my subjectivity as the resistant colonized subject (a subject moving away consciously from dominant, colonial discourses) in my pursuit to gather understandings of linguistic diversity. Initially, I related this to the country of my birth, postcolonial India, as I grew up in a society where a multilingual environment was a way of life. Later, I stumbled upon postcolonial critiques that questioned colonial ways of proclaiming invisibility and centralizing self, causing undue subordination of all the rest of the groups. I now began to contemplate and question whether my instant attachment to those theories were due to my own subjectivity, a resistant colonized subject, and I began to shy away from what I perceived to be "colonial" discourses of language. Therefore, my study almost came to a standstill as I became reluctant to even use the word "other," to name, to categorize. I even questioned this whole experience set in "colonial linguistic expression" and whether I needed to take on this linguistic identity. These expressions demanded that I take on the identity of an objective researcher who is able to distance self from topic and the study.

Memmi (1967) expresses the dilemma of the colonized, as the colonized begins to write in the language of the colonizer to address the audience (colonizer and the colonized) and to talk about the disparities created by the colonizer in the very same language of the colonizer. The language (not just English, the colonial language, but specific ways of formalizing research presentations) seemed so alien now, as it was a prescribed way of expressing what needed to be said. As a sign of resistance to colonial dominance I wanted to stop engaging in those linguistic discourses. However, my strong desire to challenge colonial dominance was rekindled when I identified the role of colonial ideologies in erasing the language histories of Australia through sociopolitical practices. This is when I earnestly wanted to reveal to current early childhood educators through my study on how we become unintentional perpetuators of dominant political discourses. I questioned whether my own subjectivity was in itself a "colonial" creation, and sometimes wanted to depose this experience in fear of submission to colonial dominance. I also realized that moving back was neither empowering nor challenging, and hence simultaneously, I questioned my own practices of being a resistant colonized subject.

Dangaremba (1988 cited in Young, 2003) talks about moving between boundaries that are layered with colonial dominance and education, as it results in a lot of uncertainty, creating an inner sense of otherness. According to Foucault, language is one of the major influences in molding our experiences, and the language discourses enable us to understand ourselves and to see the world (Danaher, Schirato, and Webb, 2000). Therefore, I began to question whether I saw myself through the language identity I had created for myself. Resistance, as being just oppositional or nonengagemental with what is colonial, doesn't cause the colonized to overcome colonial subjugation (Ashcroft, 2001). Ashcroft adds that it is only through transformation the colonized can control the future and this is about taking on the colonial discourse to transform and empower the colonized self. Therefore, I used the very same language to shift my identity without succumbing to colonial dominance and began seeing myself otherwise, as being a "resilient colonized subject."

Implications for Antiracist Pedagogy: My New Language—the Othered Talks

The colonial discourse of categorization is problematic and may only result in the eventual erosion of some languages, and the existence of one language that is dominant; and as noted earlier, ethno-linguistic groups, such as in Australia and other ethno-linguistic nation-states, are politically categorized (with the use of acronyms, such as CALD, LOTE) laden with discourses of othering and normalizing. Here in Australia, do I then refrain from categorizing, as this is perceived as a colonial act of creating iniquitous groups? My innate desire to reveal the mask of the colonizer through my study was still not extinguished. I had to group the participants involved in my study to discuss their discursive practices and how they were different or similar depending on the groups that they belonged to—an exercise that could be paralleled in many other contexts. The critical comments by Hudak (2001) and Frankenberg (1999) made me rethink about grouping. The process of identification, differentiation, and categorization involved in labeling need not always be detrimental and can be applied to objects and people (Hudak, 2001). The very act of not respectfully recognizing groups is in itself "essentialising colonial dominance" (Frankenberg, 1999). The subjective identity of a resilient colonized subject provided flexibility, as it recognized my imposed identity and yet provided me a space to challenge colonial dominance not just from the outskirts but by creating inroads to move between boundaries. I began to find ways to transform colonial discourses to unmask linguistic hegemony. Therefore, as a speaker of English as Second language, I primarily began to reframe the terms, LOTE and CALD, used to categorize "us" as the divergent members of this group "other." And I recognized that this reframing was possible across and within contexts and nation-states.

I began to use the term Languages Othered By English (LOBE) instead of Languages Other Than English to be able to distinguish the speakers of the rest of the languages that have been peripheralized by the conventional term, thereby shifting the sites of power. Moreover, this brings to notice the act of othering as being a more purposeful and deliberate act of the colonizers to assert and maintain their power and therefore not a natural phenomenon. I chose to use the term Culturally and Linguistically Identified (CALI) instead of Culturally and Linguistically Diverse, as it acknowledged only some groups being recognized. Moreover linguistically, with the removal of "diverse," this term ascribed lesser power to being normalized and peripheralized. It covertly implied that the identification is a superficial tagging by the dominant one to remain invisible. These changes made it easier for me to recognize and talk about groups and to distinguish discursive practices. I used the term "Dominant group" to denote those who positioned self as monolingual English speakers or those with English as their first language or Anglophones, including those who seemed to have adopted English as their first language without recognizing the power and privilege attached to language. I used the term "Othered group" to denote those who positioned self as nonnative English speakers, or those with English as their second language or non-Anglophones. By doing so, I was able to now challenge colonial discourses of differentiating and categorizing that covertly privileged Anglophones and silenced the speakers of LOBE. I was also able to identify the discourse of nationalizing as being instrumental in erasing the multilingual histories of Indigenous people of Australia, as English now was seen as the native language of Australia. I saw how this erasing of multilingual histories had occurred and continues to occur in many other countries that had been and are being colonized. Most of all, I was able to reveal that political ethno-linguistic grouping with embedded discourses of othering and normalizing results in early childhood educators becoming involuntary propagators of colonial dominance both here in Australia and in many other nation-states.

I am the "othered"; I talk and walk "othered," look "othered," and eat and live "othered." I realized my subjectivity as the othered only after my repeated interactions with the colonizer, who reinforced my "otherness" with all those alphabets in his language, LOTE, CALD, ESL, NESB, and what next. I now use the very same set of alphabets, LOBE and CALD, to unmask the colonizer's covert imposition and the propagation of the superior self in Australia's postcolonial era. I categorize ethno-linguistic groups to reveal colonial supremacy that aims to erase histories through discourses of nationalism. I urge individuals who are committed to challenging acts of dominance, from both "dominant" and "othered" groups, to transform subjective identities, as "resistance through resilience" can transform "silent dominant" and "silenced othered" to empowered voices. This "othered" talks now, in "Not Other Than English" (NOTE). Note this colonizer.

CHAPTER 10

Working Within, Beyond, and Through the Divides: Hopes and Possibilities for De-"Racing" Early Childhood

Sue Atkinson, Merlyne Cruz, Prasanna Srinivasan, Karina Davis, and Glenda Mac Naughton

Introduction

In this chapter we (the authors in this edited volume) come together to discuss our hopes and what we see as the possibilities for de-"racing" early childhood. We want to honor the diverse approaches and subjectivities we bring to working for the de-"racing" of early childhood without the demand and expectation of a consensus of opinions and ways forward in this work. We talk across and within our experiences as researchers (details of projects mentioned are in appendix), as educators of young children, as educators of preservice early childhood educators, and as agentic and subjective citizens. This chapter is loosely structured around our responses to each others' chapters in the book and key themes that this process raised for us. We talk about:

- Thinking about researching children, "race," and racism
- Engaging with racism
- Resistance to antiracism
- Engaging with "color-blindness"
- The disadvantages of "feeling comfortable"

- Studying white theories
- Problems with education?

We hope you find the reading of this chapter as enjoyable as it was for us in weaving it together.

Thinking about Researching Children, "Race," and Racism

Karina: [In your research project] Was your construction of early childhood and "race" different to those discussed by the participants? If so, how did this influence you during the collection and analysis of your data?

Sue: I began by thinking that young children can employ colonial constructions of "race"; and that while white children employ them to marginalize others, Indigenous children employ them to devalue themselves. I explored this idea in my interviews, as I believe that the dynamics of "race" operate in early childhood spaces. I didn't expect many of the participants to position children in this way, especially the non-Indigenous participants. I found that many participants believe that children are innocent or ignorant of "race"—even as they describe children's active understandings around "race"—and they acknowledged that these understandings emerge during the early years. I thought that those tensions in understanding could open up a space for debate around children and "race," so that issues of "race" in the early years can be addressed more actively.

Karina: How would you open that debate?

Sue: I've just finished marking about eighty assignments from a subject called The Integrated Curriculum (for an Early Childhood Education Bachelor degree course) that looked at children and "race." At the beginning of the subject, a lot of the students positioned children around ignorance and innocence, but the overwhelming majority changed their position as a result of their critical engagement with the subject's literature and workshops. They said that even though they were really surprised, shocked, and horrified that this happened, they accept that it does. They're starting to think actively about how they will behave once they graduate as early childhood educators. For example, what sorts of images and materials could they provide that would confront racism among young children? However, they also realized that just having those images and materials around them wasn't enough—they had to engage the children in a dialogue around "race" and racism. So this group of students seems to be responding quite proactively to children's knowledge of "race" and the need to challenge children's racial stereotypes in early childhood centers.

Glenda: I have found over many years of working with preservice students that the discourse of children's "racial" innocence is really prevalent. Learning that researchers over several decades have pointed to children's awareness of "race" beginning from as early as two years of age is surprising for many. However, every year several students argue that this research is not relevant in the Australian context because the majority of the studies have been U.S.-based, and Australia is not as "racially" divided as the United States. It seems to be much more powerful when they engage with local research—of which we have only a little.

Karina: Sue, did the preservice students identify what helped them?

Sue: It was some of the research, I think. For example, there was a research report about a Taiwanese child who preferred to play with white-skinned Barbie dolls. The students saw that the child was probably devaluing her own "race." Also, there was some discussion around Indigenous children, and students decided that they really needed to lift their game, because they don't know anything about their Indigenous community. Within the workshops, there were also some examples of children confronting racism.

Karina: So maybe it was a combination of the research literature and then seeing children engage with the issues?

Sue: Yes, but I'd say that a couple of them are still "in transition."

Karina: Aren't we all!

Sue: Most of them engaged actively with the idea of working with children to challenge colonial images of "race." However, some of them said that what's taught here isn't reinforced in the field. There are probably educators in the field whose ideas about children and "race" needs challenging. How do we communicate with them? When you go into an early childhood center or a kindergarten, how do you take on that role as an educator?

Karina: Sue, how could other researchers take your work further?

Sue: My research didn't involve children directly, it was only what families and early childhood educators reported back to me. I'd like to see some further research that engages directly with children—especially with Indigenous children—in early childhood spaces to explore their experiences and understandings of "race."

Glenda: It seems really important to generate that work with Indigenous children living in different contexts and places in Australia. Sue, perhaps a postdoctoral research project next?

Engaging with Racism

Sue: Why is it important to white children's wellbeing to challenge bias? Why would someone who has power and privilege and the pleasure that goes with it, want to share it?

Karina: I think that many white children in Australia live in communities that are described as multicultural, but which consist predominantly of white Anglo-Australians. Consequently, they're unprepared for racial and cultural differences. I think that we're failing these white children, because we're not giving them the skills and confidence to engage with people who look different from them. For many of them, difference is puzzling and even scary. In response, they sometimes develop a sense of superiority, a sense that their self is the one true self. Again, I think that we're failing these children, because such a sense of racial superiority prevents children from seeing themselves as part of a larger community. This means that they struggle to connect with "others" (whoever those "others" are) and act in racist ways, without understanding that that's what they're doing. So their developing sense of racial superiority isolates them from difference and this damages their sense of self, their connection to people.

Glenda: I also think that many young children have a strong sense of fairness in their daily lives. They understand the power of others to do and say things that seem unfair to them. It seems critical to build a sense of "racial" fairness and unfairness if we are to find ways to live together in "racially" just ways. Building the desire to do this in our youngest citizens seems to me a statement of hope for our future. I think hope is important to us all.

Sue: Karina, where do you think class enters into this debate about racism? If an Anglo-Saxon working-class person feels disempowered, or inferior, do you think that they empower themselves by disempowering other people? Do you think that we need to empower working-class children, regardless of their "race"?

Karina: I think so, because within white communities, racism manifests itself in different ways in different classes. For example, "race" is sometimes invoked as an explanation for unemployment and immigrants are seen as a threat to "the Australian way of life." I think that it's important to show that there are great similarities between people of the same class, regardless of their "race." If we can start to unpack the fear around job security and economics, we can show people that there are connections within a class and across "race"s that can counter that fear. We can show that they can, in fact, challenge a system that works against different groups' interests in different ways.

Prasanna: Some recent research (N/a, 2008) found that New South Wales had the most racist attitudes in Australia because they have the most migrants.

Sue: There is an expectation that when more people are mixing together, it fuels racism.

Karina: Well, it brings it to the surface.

Glenda: And, that talking about it makes it worse and breeds racism. There is real confusion about what racism is.

Sue: Yes, because people get frightened about it.

Merlyne: In my doctoral study, what triggered the interest in diversity among early childhood educators around the world is—diversity is always linked to oppression and inequity and, therefore to an interest in social justice. As my participants observed, work on diversity, oppression, and inequity around the world is focused on the marginalized and disadvantaged, that is, people from outside the dominant culture. Diversity is linked to our perceptions of difference and our attitudes to it; and racism is quite systemic. For example, immigration policies put boundaries—literal and perceptual—between different peoples.

Karina: So the people in your study said that they want to work on cultural diversity because it is linked with oppression?

Merlyne: That's right. You can't talk about diversity without talking about social justice. Racism comes in different shades. Now we say that we've "progressed," but racism persists, but in subtle modes.

Karina: So how would early childhood people talk about racism in their context?

Merlyne: Well the people in my study said that they became interested in diversity after they saw so many families marginalized because their communities lacked experience with difference.

Karina: How do people come to see that children and racism is interesting and a problem?

Merlyne: For most of them it's a journey. They start by seeing oppression and silencing; but also seeing beauty in the relationships among families that they have not seen in their communities. They see the sense of community and the beauty of those communities' values. They see the belief that you *can* do something, that no one has the right to oppress anyone. And they talk about hope and the spirituality of moving to a higher plane and seeing what is and what ought to be, and reflecting on how do we get from here to there, even though there's a sense of hopelessness.

Karina: Can we ever "be" beyond "race"?

Merlyne: I think "maybe." It's hard to say "yes" when—like I said before—"race" takes on a subtler disguise. But we can see beyond "race"—that's why we engage in antiracist education. As long as we believe in human agency and we have hope, we can be optimistic. "Race" is not a fact of nature, it's a socially constructed means of classifying, for creating so-called order in society. However, "race" has multiple meanings. Just as you can learn its dominant meanings, you can also unlearn them, contest them, and create new ones that offer a basis for social policies to promote equity and justice. At first, I found it hard to map the relations of power around "race," because I was not used to seeing it that way. Now, I "see" "race" through anticolonial frameworks. I think as long as the colonized have the courage and the strength to free themselves from being colonized, oppressed, or dominated, we can rise above our present position.

Resistance to Antiracism

Prasanna: I am more attuned to oppression because I am an Indian Brahman. I have been the oppressor, I know what oppression can do to societies and I have been made aware from a very young age what Brahmans have done. So histories were kept alive and they always told stories of creating oppression. But we don't want to create the present as the history of the future, so if we change the present, we change the future. When I came to Australia, I couldn't *not* see the oppression, I couldn't *not* see the histories being lost. That's one of the main reasons that I explore differences with students.

Karina: Prasanna, we've talked about preservice students' resistance to engaging with racism and colonialism, what are the greatest barriers they're raising?

Prasanna: The first one is an accusation: "You are racist, that's why you're calling us white!" I'm pointing out differences, but they want to be in a society that overlooks differences, so immediately they accuse me and say that I must be a very angry person to think that there are oppressed groups. Students' second response to statements about racism and oppression is to assert that they've changed under the influence of other cultures. They have embraced differences. For instance, they dine in an Italian restaurant or they have a lovely curry. So they say, "You're trying to make up stories, because they are not here. I buy curry powder, after all!"

Karina: What sorts of students are raising those barriers? Our student body is still diverse, although that diversity is decreasing.

Prasanna: Usually, it is the Anglo students, but some are from marginalized groups, such as Greeks and Italians. They say, "You are the one bringing up these things; they are not here in our society. Why should we even talk about these things? I am Italian and my parents have always said how lucky we are in this country. This country has given us so much."

I understand that view, perhaps because, since I was privileged in my country, I did not have those feelings when I came here. I thought, "This country is giving me all these things and places that I have never envisaged before." However, as we move on with the lesson and talk about racism in children, the very same students—the Greeks and the Italians—stand up and say, "Do you mean to say that it's alright for me to say that I was called a wog and that my parents were hit in the streets in those days, so it is alright for me to talk about those things?" These were the very same students who rebuked me when I said this country has feelings of racism. They were the ones who said that their parents have told them that they were the lucky ones. Now, they feel that they can't say that they've been unlucky at times and that this country has allowed that to happen. That has been very different experiences for them.

Sue: Prasanna, do you think that your presence has encouraged Asian students to identify themselves as Asian?

Prasanna: Definitely. Some of the assignments ask them to explain why they think it is important to teach children particular things, or to teach them in a particular way. I always tell them: "I believe passionately that I have to talk about these things, because of my experiences and what I am. This subject's assessment offers you a chance to express all those feelings and expressions." Many white students say that they don't want to do it, the assessment is too difficult and why should they have to do it, anyway. In response, I talk about my experiences as an international student, and I tell them that these sorts of experiences inevitably affect their teaching. In contrast, students from China, Vietnam, Indonesia, and the Philippines are very eager to do this assessment and they think it's a very simple one, because they've got so much to write about.

Karina: If you silence discussions of "race," you're silencing discussions of incidents and experiences of racism. Has this happened in your classes?

Prasanna: Yes. Sometimes I will say something and a white student will respond, "We would never say anything like that in the supermarket!" Then a speaker of a language that was not English will say, "Perhaps you haven't said anything like that, but I know what I have done as a young child when my parents talked in their home language in the supermarket. I have purposely moved away from them, run away somewhere, because I was so ashamed to hear my parents speak in my own language. Now I feel ashamed that I behaved like that, because I lost my language." So for various reasons, I can hear voices that have been silenced; and when I do I think, "This is the only way that we have to function in Australia, and it is alright for us to think that, and those words about it."

Sue: I think that the parents of those students from the south of Europe (e.g., Greece and Italy) were forced to give up their culture and to assimilate. Perhaps this is why subsequent generations don't see themselves as affected by racism. They don't experience racism because they act like Anglo-Saxons. They're not seen as "raced," as their grandparents or parents were.

Glenda: There is such a pervasive discourse pushing the silencing of talking about "racism" and "race" in Australia. I despair at times when I hear students and colleagues say, "But, there's no racism here!" I have heard it so often. Yet, as Sue's, Merlyne's, and Prasanna's research so powerfully shows it is a daily lived experience for many Australians. I think what would make us stronger as a nation is facing this reality and working together to change it.

Engaging with "Color-Blindness"

Prasanna: Difference seems taboo, so many people find it very difficult to talk about differences. A lot of people are "color-blind." They say, "All the children are the same. We don't look at their cultural or racial backgrounds. We just see each child as an individual." I think the first step toward change is to say, "It's alright to talk about differences and power. What do differences (e.g., racial differences) mean? How can they create disparities in power?"

Sue: There's a widespread belief in the early childhood field that if we talk about "race" and color, it will make children aware of "race" and, therefore, encourage them to be racist.

Karina: Sue, in your study, you found a widespread belief that children are innocent of "race" and of racism and that it's only adults who are concerned with it. And so adults ignore the effect that "race" and racism has on all children—white children as well as the children who are experiencing racism.

Merlyne: That "color-blindness" also leads people to say, "Let's all be nice and only talk about similarities. Talking about differences only creates tensions and challenges." I've devised some topics about diversity for preservice students to talk about—first in pairs, then in small groups, then as a whole group. First, I asked them to talk about one good thing about their culture. They liked that and talked about, for example, food and family values. Second, I asked them to talk about the challenges that they'd experienced around diversity. They liked that too, and became very engaged in it. Next, I asked them to talk about their experience of being an "other." (Whether they're white or not, they've all been stereotyped as an "other" in some way.) And they said, "We never get a chance to really talk about this. We're not given a lot of opportunities to discuss these things."

The challenge, of course, is to avoid making individuals feel guilty. But some white students become upset with this, it makes them uncomfortable and they want to avoid feeling uncomfortable.

For their final discussion, I asked them whether they knew of occasions when their cultural group (or any social group they belong to) had overcome obstacles and hurdles. I wanted to make the students aware of the history and context of contemporary issues around diversity. Most of them talked about their parents who had experienced discrimination; and they said that the feeling now is that "They went through it the hard way so that we can have the benefits and go to university and all that."

It was a good way to open up discussions, I think. It offered a safe space—I would call it the decolonizing of space. Hopefully, it enables you to move from one stage to the next in that journey, getting more equitable as you go.

Glenda: I like the idea of calling those discussions the decolonizing of space. I think it would be interesting to have the time and means to talk with students and others in the early childhood field about what it takes to create the safe space to do this decolonizing work.

The Disadvantages of "Feeling Comfortable"

Prasanna: When you talked about decolonizing space, sometimes in my country the colonizer has to be put in the colonized space, to be colonized. In other words the colonizer can never be colonized. They have to be colonized to decolonize.

Karina: Is it necessary to feel uncomfortable to become engaged? Imagine a life when you've never felt uncomfortable—like for many white individuals. That privileged position brings some extraordinary benefits, but how much do they miss?

Merlyne: That feeling of uncomfortableness can make a white individual feel very isolated. I think that it's really important that white people feel that, because it happens to people every day. How do we get them to think about how—as early childhood educators—they will welcome and greet the diverse families and children with whom they will work?

Glenda: Yet, there is growing research evidence that discomfort and guilt may be necessary and helpful to moving white people to action.

Sue: It will be interesting to see whether the feelings of uncomfortableness helps for white individuals to decide that as an educator they should challenge racism or whether they just ignore or dismiss the question.

Prasanna: My daughter said to me one day: "Mum, they want us to be like them and we think we are becoming like them, we take on their language and we take on their clothes and their way of life, and we walk about their society. But they never forget to remind you that you are not them, by asking you again and again, 'Where are you from?' You think you have made that many changes in yourself in order to be like them, but you will never be one of them."

Karina: If "they" need to be like "you" but they never can, how can we build early childhood communities that are decolonized spaces founded on social justice and equity?

Prasanna: I feel that this cold Australian identity has to be reconstructed on the basis of our multiplicity. New images, new discourses, new ways to speak—in every manner possible. It needs to be seen, it needs to be heard, it needs to be practiced over and over again, in all those political and social institutions. Only then, I think, will the dominant identity slowly become diluted.

Glenda: Do you ever glimpse moments when that happens? Are there points of hope?

Karina: What's that song they sing on the radio? "I am, you are, we are Australian." I've heard it time and time again. Is that really how people in early childhood communities—generally white, formally educated people—think? They're so overwhelmed and threatened by the shifting of their power base.

Merlyne: Which brings us back to students' assertions that they've changed under the influence of other cultures. Early childhood educators like to be engaged, and so to teach children about "race" as a social construction, you have to start with the educators and deal with their personal inclinations and all that. Exposing them to multiplicities is a way to make them remeet history, including the history in ourselves and where our histories position us. When I tell my personal story, the usual reaction is silence, or "That made me think." One white preservice student said, "That's your story, but I don't care, it doesn't affect me. I'll never be like the person you're talking about." I replied that while this is the story of my life, it can still affect you because each of us affects other people's stories by our stories. She said, "No. It's *your* story. It's nothing to do with me."

Prasanna: Sometimes I wonder: what if every politician were made to learn an Indigenous language? How "uncomfortable" they would feel. How disempowered they would feel, and how empowered every Indigenous language speaker would feel. The colonizer would have to learn from the colonized.

Merlyne: I've been a volunteer for Australian Multicultural Education Services. The first exercise that they gave us was to fill in a form in another language. My first reaction was, "What am I going to do?" Yet that's what we do all the time: we say, "Here you go, work it out."

Karina: And you're going to be assessed on this.

Merlyne: That's right. So I think that educators should study white as a theoretical understanding of "race."

Studying White Theory

Karina: How do you present white as a theoretical understanding of "race" without rearticulating white dominance?

Prasanna: White and the white culture permeates every aspect of the society, but there's no recognition and acknowledgment from many white people, so that's why that's important.

Karina: When I talk about experiences of being white, white people often reply that they're insulted: "What are you talking about? How can you say that?" It's important to say that many white people don't see themselves as racially or culturally biased, so they need to hear from people who face racism, they need to know what's going on. For example, the people who respond positively to my remarks about white identities have nearly always faced racism. To them, white and its discourses are quite clearly the issue: they say to me, "Well, of course that happens!"

Merlyne: I think it's quite important to hear powerful narratives from white people who have reflected critically on their position as white people and can, as a result, engage in conversations like this. I think that such stories can show how "being comfortable" can lead to a poor sense of identity, both for children and for adults. How enriching it is to be immersed in multiplicities and in diversity! Those stories can also expose the effect of white both on white people and on "nonwhite" people.

Sue: Karina, you did that, didn't you? I think it was a couple of years ago. You were being quite confrontational with your students and they resisted. You diverted the conversation to an ethics based one about justice and that tended to shift people more strongly.

Karina: I said to white students, "You're white like me, and you need to realize how bad that is!" I felt it was really important to do that and I still think at times that it is really important to do it. At the same time, I found in Foucault's (1994) work and in some of Jennifer Gore's (1993) work the idea of an ethics of care, and of care of the self. You can't be ethical unless you are concerned about the issues of ethics and equity in the lives of other people, and whether your interactions with them might be inequitable. That opens a different space, I think.

Glenda: It engages students with the politics of ethics in their daily lives. I know that has been a helpful concept to use in work with educators on these issues. How do you care for yourself and care for the other amongst our differences and similarities can be challenging to ask of educators. How do you welcome "the other" in your daily practice with children, colleagues, and parents?

Problems with Education?

Sue: We're assuming that early childhood educators—and maybe many of our students—strive to be ethical. But when it comes to "race" and gender, you need to talk about rights and power, and they're the difficult things. Few students think of the education system—whether it's early childhood or secondary—as institutionalized and oppressive. How do you get them to see it?

Karina: Often I ask them what they see as their role in education? Do they think that education should promote assimilation?

Sue: There are probably some people who think that that's not their role.

Karina: Yes, but others are often caught in pedagogical structures that are so dominant that they are hard to see past. Nonetheless, they ask themselves, "What do I want to do? What is my purpose in being an educator?" If you encourage them to do that, then they can see a mismatch between their aims and what they're caught up in doing. And then you offer them multiple ways of being an educator.

Prasanna: The whole society upholds the white way to be and to practice, so how will white educators ever feel a need to embrace "the other"?

Sue: White people have greater access to power, but for some that's not enough. They want a more ethical, spiritual way of life, connecting across culture, and they want to pull down racism and to build power in a different way in their community.

Glenda: It's challenging however to learn how to exercise my power as a white person to do that. I can't deny the power my white identity brings, but asking how I (and other white Australians) might use that power with antiracist effect seems critical to pulling down racism.

Karina: Doesn't that sound exciting? If we approach the task of fighting racism in that way, we will force these structures to shift, because the mismatch between what people want and the operation of those pedagogical structures would be too great. If early childhood educators want to be seen to be doing their job well and providing a high-quality service (however that's defined), they have to have access to what the child is saying all the time, because they're meant to assess it. They have to have access to what the families think about the child, because they're meant to assess that, too. That environment is so racialized and colonizing that we need to deconstruct it if we're gong to create a setting of multiplicity that we can all share.

Merlyne: Learning to listen and look at things in multiple ways is exciting. Like reading in multiple ways—from the angle of the child, of the parents, and so on.

Prasanna: I don't want to decolonize the colonizer, I want to decolonize myself, and colonize the colonizer!

Merlyne: I think an anticolonial approach to teaching moves you away from discussing oppression and toward discussing agency, community, and spirituality and all of that, which are all, I think, positive frameworks of culture and identity. As Karina said, it's very exciting. For us anyway! We don't know what it means to others!

Prasanna: My trouble or struggle has always been that the colonizer easily moves around and claims ownership of everything.

Sue: So just because you now eat curry, it doesn't make you a good person!

Merlyne: Resistance or resilience is always part of anticolonialism, but what does that mean? Is it being nice? How do you move beyond resistance?

Prasanna: If you are part of what has happened, you are never empowered to make those changes. You resist and you move away and you think, "What is colonial? Something that needs to be deposed." You need to be cleansed. I have to think that it is OK for me to take on certain aspects of the colonizer, but with specific motivations. I know what I am doing. I am doing it not because I feel inferior, but because I am empowered.

Merlyne: So it's not running away from it, it's going through it.

Prasanna: It's resistance with a definite motive. I can never be immunized. I can never think outside of resistance, but to resist colonialism is not necessarily to move away from it. It is to come back to it. To stand strong in it without feelings of superiority or inferiority.

Part II References

Alegre, E.N. (1993). *Pinoy forever: Essays on culture and language.* Pasig, Metro Manila: Anvil.

Armstrong, J., and R. Ng. (2005). A conversation between Jeanette Armstrong and Roxanna Ng. Deconstructing race, deconstructing racism: Critical essays for activists and scholars. In J.A. Lee and J. Lutz (Eds). *Situating "race" and racisms in space, time, and theory* (pp. 30–45). Montreal, London: McGill Queen's University Press.

Arthur, L. (2001). Diverse languages and dialects. In E. Dau (Ed). *The anti-bias approach in early childhood* (pp. 95–113). NSW: Pearson Education Australia.

Asante, M. (2006). Foreword. In G.S. Dei and A. Kempf (Eds). *Anti-colonialism and education: politics of resistance* (pp. ix–x). Rotterdam, The Netherlands: Sense.

Ashcroft, B. (2001). *On post-colonial futures: transformations of colonial culture.* New York: Continuum.

Berlak, A. and S. Moyenda. (2007). Teaching story: Reflections upon racism and schooling from kindergarten to college. In K.K. Kumashiro and B. Ngo (Eds). *Six lenses for anti-oppressive education* (pp. 193–210). New York: Peter Lang.

Burnett, B. (2001). Coming to terms with culture and racism. *Contemporary Issues in Early Childhood*, 2(1), 105–108.

Cannella, G.S. (1997). *Deconstructing early childhood education. Social justice and revolution.* New York: Peter Lang.

Clyne, M. (2005). *Australia's language potential.* NSW: University of NSW Press.

Commonwealth of Australia. (2006). *Australian citizenship. Much more than a ceremony.* Canberra: AGPS.

Constantino, R. (1966). *The miseducation of the Filipino.* Quezon City: Foundation for Nationalist Studies.

——— (1975). Our captive minds. In Pilipinos in American History workgroup (Eds). *Iriri Ti Pagsayaatan Ti Sapasap: A reader on the history of Pilipinos in America.* San Francisco: Pilipinos in American History workgroup.

Corson, D. (1993). *Language, minority education and gender: Linking social justice and power.* Clevedon: Multilingual Matters.

Danaher, G., T. Schirato, and J. Webb. (2000). *Understanding Foucault.* NSW: Allen and Unwin.

Davis, K. (2005). Troubling practice: Exploring white dichotomies in early childhood practice. *The International Journal of Equity and Innovation*, 3(1), 18–30.

De Mesa, J. (1987). *In solidarity with the culture: Studies in theological re-rooting*. Quezon City, Philippines: Maryhill School of Theology.

Dei, G.S. (1999). Knowledge and the politics of social change: The implication of anti-racism. *British Journal of Sociology of Education*, 20(3), 395–409.

———. (2006). *Anti-colonialism and education: The politics of resistance*. Rotterdam, The Netherlands: Sense.

Delgado, R. and J. Stefancic. Eds. (1997). *Critical white studies. Looking behind the mirror*. Philadelphia: Temple University Press.

Department of the Prime Minister and Cabinet. Office of Multicultural Affairs. (1999). *A new agenda for multicultural Australia*. Canberra: AGPS.

Derman-Sparks, L. (1989). *Anti-Bias curriculum: Tools for empowering young children*. Washington: National Association for the Education of Young Children.

Derrida, J. (1973). *Speech and phenomena*. Evanston, IL: Northwestern University Press.

——— (1981) *Positions*. Cited in Grosz, E. 1990. Contemporary theories of power and subjectivity. In S. Gunew (Ed). *Feminist knowledge critique and construct*. London: Routledge.

Dudgeon, P., and D. Oxenham. (1990). *The complexity of Aboriginal diversity and kindredness*. Aboriginal and Torres Strait Islander Unit, University of Queensland: Queensland.

Elder, C. (2005). Immigration history. In M. Lyons and P. Russell (Eds). *Australia's history: Themes and debates* (pp. 98–115). Sydney: UNSW Press.

Enriquez, V. (1986). *Philippine worldview*. Singapore: Institute of Southeast Asian Studies.

Fanon, F. (1963). *The wretched of the earth*. Trans. Constance Farrington (1963 translation of the 1961 book) New York, Grove Weidenfeld.

———. (1967). *Black skin, white masks*. New York: Grove Press.

Feagin (2006). *Systemic racism, a theory of oppression*. New York: Routledge.

Fine, M., L. Weis, L.C. Powell, and L.M. Wong. (1997). Preface. In Fine, et al. (Eds). *Off White: Readings on race, power and society* (pp. vi–xii). New York: Routledge.

Fleer, M., and B. Raban. (2005) *"It's the thought that counts": A sociocultural framework for supporting early literacy and numeracy*. Paper presented at Our Children the Future 4, Adelaide, June.

Foucault, M. (1994). *The essential works of Michel Foucault. Ethics, subjectivity and truth V1*. Trans. R. Hurley et al. New York: New Press.

Frankenberg, R. (1993). *White women, race matters. The social construction of white*. Minneapolis, MN: University of Minnesota Press.

———. *The social construction of white women, whiteness race matters.* Minneapolis: University of Minnesota Press.

Freire, P. (2001). *Pedagogy of the oppressed.* New York: Continuum.

Gallini, C. (1996). Mass exoticisms. In I. Chambers and L. Curti (Eds). *The Post-colonial question: Common skies, divided horizons.* London: Routledge.

Gandhi, L. (1998). *Postcolonial theory: A critical introduction.* New York: Columbia University Press.

Goldberg, D. (1990). *Anatomy of racism.* Minneapolis, MN: University of Minnesota Press.

———. (1993). *Racist culture: philosophy and the politics of meaning.* Massachusetts: Blackwell.

Gore, J.M. (1993). *The struggle for pedagogies. Critical and feminist discourses as regimes of truth.* New York: Routledge.

Gramsci, A. (1971). *Selections from the prison notebooks.* New York: International Publishers.

Grin, F. (2005). Linguistic human rights as a source of policy guidelines: A critical assessment. *Journal of Sociolinguistics*, 9(3), 448–460.

Hage, G. (2000). *White nation. Fantasies of white supremacy in a multicultural society.* Annandale, NSW: Routledge.

Hall, S. (1997). The spectacle of the "other." In S. Hall (Ed). *Representation: Cultural representations and signifying practices* (pp. 223–290). London: Sage.

Harrison, F. (2002). Unravelling race for the twenty-first century. In J. Mac Clancy (Ed). *Exotic no more* (pp. 145–166). Chicago: University of Chicago Press.

hooks, b. (2002). *Rock my soul: black people and self-esteem.* New York: Washington Square Press.

Howard, G.R. (1999). *We can't teach what we don't know. White teachers, multiracial schools.* New York: Teachers College Press.

Hudak, G.M. (2001). On what is labelled "playing". Locating the "true" in education. In G.M. Hudak and P. Kihn (Eds). *Labelling: Pedagogy and politics* (pp. 9–26). London: Routledge-Falmer.

Joliffe, J. (2004). The language of learning. *The Age (Education)*, October 25, p. 4.

Joseph, C., J. Lane, and S. Sharma. (1994). No equality, no quality. In P. Moss and A. Pence (Eds). *Valuing quality in early childhood services: New approaches to defining quality.* New York: Teachers College Press.

Kailin, J. (2002). *Antiracist education: From theory to practice.* Philadelphia, PA: Rowland and Littlefield.

Kane, N. (2007). Frantz Fanon's theory of racialization: Implications for globalization. *Human Architecture Journal of the Sociology of Self-knowledge* 5(Summer), 353–362.

Kelen, C. (2005). Hymns for and from white Australia. In A.J. Lopez (Ed). *Postcolonial white: A critical reader on race and empire* (pp. 201–230). Albany, New York: State University of New York Press.

Ladson-Billings, G., and W.F. Tate. (1995). Toward a critical race theory of education. *Teachers College Record*, 97(1), 47–68.

Loomba, A. (1998). *Colonialism/post colonialism.* London, Routledge.

Mac Naughton, G. (2000). *Blushes and birthday parties: Telling silences in young children's construction of "race."* Paper presented at the Australian Research in Early Childhood Education Annual Conference, January 29–30, Canberra.

———. (2001). Silences, sex-roles and subjectivities: 40 years of gender in the Australian journal of early childhood. *Australian Journal of Early Childhood*, 26(1), 21–28.

———. *Doing Foucault in early childhood studies: Applying post structural ideas.* London and New York: Routledge Falmer.

Mac Naughton, G., and K. Davis. (2001). Beyond "Othering": Rethinking approaches to teaching young Anglo-Saxon children about Indigenous Australians. *Contemporary Issues in Early Childhood*, 2(1), 83–93.

Makin, L., J. Campbell, and C. Jones Diaz. (1995). *One childhood, many languages.* NSW: Harper Educational.

Martinot, S. (2003). *The rule of racialization: class, identity, governance.* Philadelphia: Temple University Press.

McLaren, P. (1997). *Revolutionary multiculturalism. Pedagogies of dissent for the new millennium.* Colorado: Westview Press.

McLaren, P., and P. Leonard. (1993). *Paulo Freire: A critical encounter.* New York: Routledge.

McLaren, P., and T. Da Silva. (1993). Decentering pedagogy: Critical literacy, resistance and the politics of memory. In P. McLaren and P. Leonard (Eds). *Paulo Freire: A critical encounter* (pp. 47–89). New York: Routledge.

Memmi, A. (1967). *The colonizer and the colonized.* Massachusetts: Beacon Press.

———. (2000). *Racism.* Trans. Steve Martinot. Minneapolis, MN: University of Minnesota Press.

Miles, R. (1982). *Racism and migrant labour.* London: Routledge.

Mills, S. (2004). *Discourse.* London: Routledge Taylor and Francis.

Moran, A. (2005). *Australia: Nation, belonging and globalization.* New York: Taylor and Francis.

Moreton-Robinson, A. (2000). Duggaibah or "place of white": Australian feminists and race. In J. Docker and G. Fischer (Eds). *Race, color and identity in Australia and New Zealand* (pp. 240–255). Sydney: University of New South Wales Press.

———. Ed. (2004). *Whitening race: Essays in social and cultural criticism.* Canberra: Aboriginal Studies Press.

Morris, B., and G. Cowlishaw. (1997). Cultural racism. In G. Cowlishaw and B. Morris (Eds). *Race matters. Indigenous Australians and "our" society* (pp. 1–10). Canberra: Aboriginal Studies Press.

N/a, (2008). 40pc believe others don't belong here. *The Australian*, October 21. Sourced at http://www.theaustralian.news.com.au/story/0,25197,24415273-12377,00.html on November 1.

Nai-Lin Chang, H., and D. Pulido. (1994). *The critical importance of cultural and linguistic continuity for infants and toddlers.* (Online). Available: http://www.californiatomorrow.org/files/pdfs/Cultural-Linguistic-Continuity.pdf accessed on August 28, 2008.

National Childcare Accreditation Council (NCAC). (2001). *QIAS Source book extract.* (Online) Available: http://www.ncac.gov.au/previous_qias_source_book_preamble.pdf.

Neumann, K. (2004). *Refuge Australia. Australia's humanitarian record.* Sydney: University of New South Wales Press.

Omi, M., and H. Winant. (1986). *Racial formation in the United States from the 1960's–1980's.* New York: Routledge.

———. (1994). *Racial formation in the United States.* New York, Routledge.

Pierce, L. (2005). Not just my closet: Exposing familial, cultural, and imperial skeletons. In M. de Jesus (Ed.) *Pinay power: Theorizing the Filipina/American experience.* New York: Routledge.

Quezon, M. (1966). Philippine racism. *Philippine Graphic*, November 2.

Raban, B., and C. Ure. (2000). Cross-national perspectives on empowerment—a guide for the new millennium. In J. Hayden (Ed). *Landscapes in early childhood education* (pp. 375–390). New York: Peter Lang.

Rafael, V. (2000). *White love and other events in Filipino history.* Durham NC: Duke University Press.

Rimonte, N. (1997). Colonialism's legacy: The inferiorizing of the Filipino. In M. Root (Ed). *Filipino Americans, transformation and identity* (pp. 39–61). Thousand Oaks, CA: Sage.

Rizal, J. (1889/1990). Rizal's letter to Blumentritt. *The Rizal-Blumentritt Correspondence*, 2(1), 117–118.

San Juan, E. (1994). The predicament of Filipinos in the United States: Where are you from? When are you going back? In K.A. San Juan (Ed). *The state of Asian America: Activism and resistance in the 1990's* (pp. 205–214). Boston: South End Press.

Russell, P. (2005). Unsettling settler society. In M. Lyons and P. Russell (Eds). *Australia's history: Themes and debates* (pp. 22–40). Sydney: University of New South Wales Press.

Said, E.W. (1978). *Orientalism: Western conceptions of the orient.* London: Penguin.

Siraj-Blatchford, I., and P. Clarke. (2000). *Supporting identity, diversity and language in early years.* Buckingham: Open University Press.

Skattebol, J. (2005). Insider/outsider belongings: Traversing the borders of whiteness in early childhood. *Contemporary Issues in Early Childhood*, 6(2), 189–203.

Skutnabb-Kangas, T. (1988). Multilingualism and the education of minority children. In T. Skutnabb-Kangas and J. Cummins (Eds). *Minority education: From shame to struggle* (pp. 9–44). Clevedon: Multilingual Matters.

Spencer, E. (2006). Spiritual politics: Politicizing the black church tradition in anti-colonial praxis. In G.S. Dei and A. Kempf (Eds). *Anti-colonialism and education: The politics of resistance* (pp. 107–128). Rotterdam, The Netherlands: Sense.

Strobel, L.M. (2000). Born-again Filipino. Filipino American identity and Asian panethnicity. In A. Singh and P. Schmidt (Eds). *Postcolonial theory and the United States: Race, ethnicity, and literature* (pp. 349–369). Jackson: University Press of Mississippi.

Taylor, A. (2005). Situating whiteness critique in Australian early childhood: The cultural politics of Aussie kids in the sandpit. *International Journal of Equity and Innovation in early childhood*, 3(1), 5–17.

The Parliament of the Commonwealth of Australia (1993). *The literacy challenge: A report on strategies for early intervention for literacy and learning for Australian children.* Canberra: AGPS.

Unsal, S. (2006). Implicit racism and the brain: How neurobiology can inform an anti-colonial, anti-racist pedagogy. In G.S. Dei and A. Kempf (Eds). *Anti-colonialism and education* (pp. 63–86). Rotterdam: The Netherlands. Sense.

Van Ausdale, D., and J. Feagin. (2001). *The first R: How children learn race/racism.* Maryland: Rowman and Littlefield.

Van Tiggelen, J. (2005). The sound of one man chatting. *Good weekend: The Age magazine*, September, pp. 24–30.

Viruru, R. (2001). Colonized through language: The case of early childhood education. *Contemporary issues in early childhood*, 2, 31–47.

Walker, Y. (1993) Aboriginal family issues. *Family Matters*, 35, 51–53.

Young, R.J.C. (2001). *Post colonialism: An historical introduction.* Oxford: Blackwell.

———. (2003). *Postcolonialism: A very short introduction.* Oxford: Oxford University Press.

APPENDIX

Research Project Descriptions

1. The Preschool Children's Constructions of Racial and Cultural Diversity (PCCRCD) Research Project (Glenda Mac Naughton)

In this Australian Research Council funded grant we undertook a three-year, field-based study to investigate how eighty-four preschool (three to five years of age) girls and boys from differing racial backgrounds constructed their cultural and "racial" understandings. This field work was conducted in two inner Melbourne early childhood services, one suburban Melbourne early childhood service, and one Victorian early childhood service located in a regional town. Two of these services were located in areas of considerable racial and cultural diversity, one service was in an area that had some racial and cultural diversity, and one service was in an area that was predominantly Anglo-white.

Three questions guided the research work within this project. These were:

- What relationships exist between preschool children's understandings of cultural and "racial" diversity and of their own gender, class, and ethnic identities?
- What factors influence preschool children's cultural and "racial" understandings over time?
- How can these relationships and understandings best be theorized?

Two researchers worked in two services each and were in the field with the children over a period of eighteen months. Across this time, data was collected in the forms of:

- Researcher observations of play and interactions
- Two individual interviews with each child participant. The interviews were structured around:
  - Using Persona Dolls to discuss and explore children's knowledge and understandings of cultural diversity and explore friendship patterns
  - Using self-portraiture and portraiture to discuss and explore identity of child and their friends
- Group discussions using Persona Dolls to explore issues raised in the research
- Individual discussions with children about their participation in the project.

Whilst we focused in this project on children's understandings of racial and cultural diversity, we were also mindful of the calls for children's voice to inform our knowledge about them (Cannella, 1997; Christensen and James, 2000; Mac Naughton and Smith, 2005). To honor these calls in this project we:

- allowed children to shape their research identities by choosing their pseudonyms
- centered children's choices during data collection by honoring their right to withdraw data from the project and at any point in time to say no to data collection
- centered children's efforts to direct research techniques by allowing their agenda to dominate interviews.

This project found that young children where aware of "race" in their lives and constructed and positioned their own and others' identities according to politicized understandings of "race."

2. Indigenous Self-Determination and Early Childhood Education and Care in Victoria (Sue Atkinson)

This was a qualitative study of how Victoria's early childhood community negotiates colonial constructions of Aboriginality around dualisms such as assimilated/traditional, Indigenous/non-Indigenous, and intersecting colonial constructions of the child as ignorant or innocent of "race" and power.

Five questions guided the study:

- What is the early childhood center's role in building on the identity and culture of Indigenous preschoolers?
- How do early childhood centers plan and implement Indigenous inclusive programs for Indigenous and non-Indigenous preschoolers?
- What understandings of Victoria's Indigenous culture do non-Indigenous early childhood practitioners hold?
- What understandings of Victoria's Indigenous culture do Indigenous and non-Indigenous preschoolers hold?
- How is Indigenous community consultation being implemented in the construction of Indigenous early childhood curricula?

The project investigated these questions by interviewing thirty-nine members of Victoria's early childhood community: Indigenous elders, Indigenous and non-Indigenous early childhood professionals, Indigenous parents, and Indigenous preschoolers. (These participants are referred to only by pseudonyms in this book in line with the University of Melbourne's Human Research Ethics Committee protocols.)

The aim of this project was to build an Indigenous early childhood curriculum that contributes to and is an expression of the self-determination of Indigenous people in Victoria. In order for such a curriculum to go forward, this project found a need for a reconceptualization of Aboriginality around complexity and multiplicity, as well as continuity and uniformity, which can better address those issues of "race," culture, identity, and racism that see Indigenous communities marginalized within non-Indigenous programs.

3A. Beliefs and Preparation about Teaching in a Multicultural Society (Merlyne Cruz)

My honors study addressed the following broad questions: 1) What beliefs do early childhood preservice educators have regarding ethnic diversity in relation to teaching and learning in early childhood settings? and 2) How do early childhood preservice educators view the effectiveness of their education program in relation to preparing them for teaching in a multicultural society? The research project revealed several key findings. The lack of quality models of culturally inclusive curriculum implemented in most of their practicum placements was one. The growing trepidation toward increased diversity in the population was another. Coupled with the latter concern was the realization that, as future educators, they have a responsibility to respond to the trend in an appropriate and respectful way. As prospective educators (most of whom have had limited exposure with ethnically diverse population in their life experiences and limited experiences with diverse children and families in their educator training), this reality seemed overwhelming. The participants in this study believed that their educational training has positively influenced their knowledge and dispositions toward diversity. However, they were also aware that they have come to their educator education programs with belief systems that have already been established. Thus, learning how to be inclusive and respectful of diversity would be an ongoing journey. There were ideological tensions that had to be dealt with. New insights about one's self and about others were being formed. There were "shifts" and "tensions" to grapple with, as one attempts to interpret new (antibias/antidiscriminatory) understandings through new lenses and replace one's existing belief system with a new worldview. My honors thesis sparked an interest in learning more from early childhood educators who were role models and leaders in inclusive practices.

3B. Early Childhood Educators and Commitments to Honoring Cultural Diversity (Merlyne Cruz)

In my current doctoral research, I explore the lived experiences of early childhood educators who are committed to honoring cultural diversity. Using an anticolonial framework, I consider various dimensions that helped create and sustain their respect for diversity. This research emerged at a transformative moment in my life when I had woken up to the dehumanizing and oppressing impact of my colonial history. Speaking from a simultaneously colonized/decolonizing/decolonized voice and taking on a vantage point from the margins, I examine early childhood educators' understandings on cultural diversity in the colonial structures of where we teach and live.

4. The Inclusion of Linguistic Diversity in Early Childhood Settings Project (Prasanna Srinivasan)

My masters thesis was set within a larger project called Respectable Equitable Staff Parent Engagement in Children's Services Today

(RESPECT) that explored the current staff-parent relationships in children's services. The main project RESPECT spanned over a period of two–three years (January 2006–December 2008), and drew from data from children's services in Australia, especially those that provide long daycare, in the following three states namely Victoria, South Australia, and New South Wales (NSW).

In my study, I used data collected by interviewing (semi-structured face-to-face interviews) ten early childhood practitioners, who worked in long daycare centers in Melbourne metropolitan region. My study originally aimed to answer the following questions:

- How is inclusion of linguistic diversity understood by early childhood practitioners?
- What are considered as practices of inclusion?
- What are the factors that influence early childhood practitioners' attitudes and practices?

I first analyzed the data using an interpretivist paradigm using theories of language and cultural identity development, and the conclusion was that most early childhood practitioners were aware of the role of languages in cultural identity development and therefore believed their practices supported children's ethno-linguistic background. However, the purpose of my study evolved with the introduction of postcolonial theoretical framework and later on I analyzed the same set of data using this framework. The study's questions now included:

- Are early childhood practitioners' understandings influenced by colonial attitudes that were already evident in Australia's political discourses?
- What are the postcolonial elements reflected in their practices?

A comparison of the two sets of analytical discussions revealed the covert dominance of colonial ideologies around languages still evidenced in Australian political discourses. Moreover, these attitudes trickled into the daily discursive practices of early childhood practitioners and thus became involuntary propagators of colonial linguistic dominance that erased the linguistic histories of Australia and asserted English as native and national identity of Australia. Thus, my study authenticated the use of postcolonial theoretical base to uncover such discourses and alert practitioners to seek alternative practices to support languages in early childhood settings.

5. Teaching for and with Cultural Diversity in Early Childhood (Karina Davis)

In this action research project funded by the University of Melbourne Early Career Researcher's Scheme, eight early childhood educators and I sought to explore our own understandings and experiences of cultural diversity and our early childhood practices through concepts drawn from postcolonial and critical white theories. We did this through:

- meeting every month across eighteen months to discuss experiences and understandings and how they intersected with practices
- exploring literature that raised issues and introduced concepts connected with postcolonial and critical race ideas and theories
- participating in one-on-one interviews both at the beginning and at the end of the project.

Using concepts from postcolonial and critical white theories made it possible to work to deconstruct the notion of the invisibility of white and provided spaces for us all to locate and explore our cultural and racial constructions. Being introduced to and hearing the words of marginalized "others" was also important in beginning to acknowledge and challenge white invisibility, dominance, and discriminatory practices. While this exploration, naming, and uncovering of white discourses was a stated aim of the research, the enacting of this was difficult in practice, as white continually sought to reposition itself as universal and dominant and we were required to see this rearticulation of white as an ongoing negotiated struggle in our work.

References

Cannella, G.S. (1997). *Deconstructing early childhood education. Social justice and revolution.* New York: Peter Lang.

Christensen, P., and A. James. Eds. (2000). *Research with children: Perspectives and practices.* New York: Routledge.

Mac Naughton, G., and K. Smith. (2005). Exploring ethics and difference: The choices and challenges of researching with children. In A. Farrell (Ed). *Exploring ethical research with children* (pp. 112–123). Buckingham: Open University Press.

Subject Index

Aboriginal, 3, 10, 11, 123, 124, 142, 144, 145, 146, 148
Aboriginal Australian, 3, 144
Aboriginality, 139, 141–142, 143–149, 151, 152, 188, 189
anticolonial, 8, 134–135, 138, 155, 171, 178, 179, 190
anticolonial thought, 135
  see also anticolonial
antiracism, 8, 133, 137, 138, 167
  feeling uncomfortable, 174
  resistance, 167
Anti-racist education
antiracist pedagogy/pedagogies, 6, 7, 8, 44, 61, 63, 64, 84, 150
attitudes, 18, 19, 20, 21, 23, 25, 26, 27, 35, 38, 40, 41, 50, 134, 145, 170, 171, 191
  race attitudes, 18, 21, 35, 38, 41, 50, 145
Australia, 2, 3, 5–7, 9–11, 25, 49, 54–61, 87, 90–92, 113, 114, 118, 121, 124, 131–134, 148, 150, 155, 159, 162, 164, 165, 169, 171, 172
  citizenship, 9, 10
  racial history, 9

black/white, 4, 26, 55, 152

care of the self, 177
chromatism, 28
cognition, 17, 20, 23, 157
Cognitive-Developmental Theory, 21–22
colonialism, 29, 54, 129, 130, 133, 134, 137, 155, 161, 172, 179
colonization, 3, 5, 9, 37, 130, 133, 134, 145, 147, 149, 152, 162
colonizing, 178
color-blind/ness, 62, 167, 174
critical engagement, 120, 122, 123, 124, 125, 168
critical race theories, 8, 36, 37, 85, 86, 88, 94, 113
critical white theories, 70, 85, 192
Culturally and Linguistically Diverse, 160, 162, 165
Culturally and Linguistically Identified, 165

decolonizing, 135
decolonization, 135, 138
dichotomy, 115, 116
discourse, 5, 9, 17, 29, 31, 33, 34–37, 42, 49, 50–54, 56–63, 67, 81, 82, 85–96, 115, 116, 118–121, 123–125, 130, 133, 134, 139, 148, 150–152, 158, 159, 161–166, 169, 173, 175, 177, 191, 192
  childhood discourses of, 139
  colonial discourses of, 9, 130, 148, 150, 162, 163, 165
  color-blindness, 62, 82
  discourses of whiteness, 34, 35, 37, 49, 52, 60, 62, 82, 85, 86, 88, 95
  masculinity/ies, 85, 86, 94, 95
  of privilege, 53, 59
  of race, 9, 17–30, 29, 34–37, 54, 56
  racial innocence of children, 39, 42, 62, 82
  white masculinity, 85

doll test, 41
  critiques, 37, 38, 40, 41, 42, 163

early childhood spaces, 7, 30, 32, 34, 36, 43, 45, 46, 47, 63, 68, 83, 95, 147, 149, 151, 152, 168–169
education
  preservice, 133, 167, 169, 172, 174, 176, 189
ethics, 130, 177, 189
ethics of care, 177
ethnographic, 31, 35, 37, 42, 43

feminist poststructuralism, 68
First Nations, 3

gender, 22, 28, 32, 35, 36, 37, 41, 50, 67–72, 77, 80–83, 85, 86, 91, 94, 95–97, 118, 117
  feminine, 68, 70, 72, 74, 75, 76, 87, 93
  white beauty ideal, 70, 71
gender race, 95
  and young boys, 85, 86, 88
  and young girls, 67, 85, 94

identity/identities, 6, 7, 19, 20, 21, 23–24, 30–34, 37, 46, 50, 61, 62, 68, 72, 76, 81, 85, 93, 99, 115–119, 121, 124, 125, 129, 130, 134–138, 146, 147, 149, 151, 152, 156, 157, 161, 163, 164, 177, 178, 189
  Aboriginal identity, 146
    see also Indigenous
  adult identity, 5, 24, 37, 86, 127
  Australian identity, 116, 175
  black identity, 57, 121, 125, 140, 141–148, 152
  children's cultural mediation of, 68
  children's identity, 24, 31, 32, 42, 50, 51, 54, 63, 67, 86, 96, 149
  colonial identity, 155
  critical engagement with, 120, 122, 123, 124, 168
  cultural identity, 115, 191
  culture identity, 72
  ethnic identity, 136
  Filipino identity, 128
  gender identity, 72
  and gender-race, 67, 68, 69, 75, 83
  identity construction, 86
  Indigenous identity, 56, 146, 152, 159
    see also Aboriginal
  language identity, 164
  masked, 113, 114–116
  naming white, 119, 124
  national identity, 117, 118, 159, 191
  non-white, 52, 56, 74, 93, 115, 118, 121, 122, 125, 132–134, 177
  politics of, 7, 30, 50, 85
  racial identity, 24, 26, 27, 32, 35, 36, 50, 72, 114, 118
  phases, 20, 27
  socially mediated, 32, 33
  racialization of, 50, 51, 62
  racialized identity, 55
  resistance identity, 81
  social identity, 26, 27
  (un)masking, 113, 114
  white Australian, 55, 59, 87, 115, 116, 117, 160, 178
  white educator, 114, 119, 124, 125, 178
  white identity, 117, 118, 119, 120, 123, 125, 178
Indigenization, 136, 136–137
Indigenous, 3, 8, 9–12, 37, 55–57, 74, 114, 123, 124, 132, 136, 139–142, 144–150, 151–152, 157, 159, 160, 161, 165, 168, 169, 176, 184, 188, 189
Indigenous Australians, 9, 105, 124, 146, 152
Intergroup Contact Assessment, 40

Languages Othered By English, 165
Languages Other Than English, 160, 165
linguicism, 8, 158

masculinity/ies, 7, 85–96
  Australian white hegemonic, 85, 89, 93
  boundary defense, 89
  challenging, 95
  definition, 86
  hegemonic, 80, 85–96, 138
  heterosexism, 70, 87, 92
  mateship, 85, 87, 92
  sanctioning of, 89, 92, 93, 94, 116, 160
migration, 49, 58, 59, 116, 132, 155
  Australia, 9–12
multiculturalism, 10, 116

normalizing, 125, 161–162, 164–165

observation, 25, 31, 32, 40, 42, 44, 61, 69, 86, 88, 89, 91, 92, 94, 95, 188
  early childhood use of, 88, 91, 92, 93
  in PCCRCD project, 87, 88, 89
  and race, 88
oppression, 2, 3, 36, 114, 125, 129, 133, 134, 135, 137, 138, 147, 171, 172, 178
  and adults, 3, 80
  and children, 36, 53
  and early childhood, 53, 114, 125
othered, 62, 71, 116, 118, 121, 165–166
othering, 36–37, 115, 139, 151, 160–162, 164–165

physical race markers, 19, 22, 23, 27, 38
popular culture icons, 7, 67, 70, 72, 75, 76, 78, 80, 81, 82,
  Barbie, 69–72, 74–76, 78, 79, 83, 165
  ethnic Barbie, 70–72
  fairies, 70, 72, 74, 75, 83
  pink, 28, 74, 76, 78, 79
  princesses, 70, 72–76, 78, 83
  proto-Barbie, 71–76
  use by children, 70
postcolonial/postcolonialism, 8, 30, 36, 37, 51, 143, 145, 148, 151, 152, 155, 159, 160, 162, 163, 166, 191, 192
postmodern/postmodernism, 8, 30, 33, 34, 37, 85, 88
poststructuralist/poststructuralism, 8, 33, 34, 36, 37, 50
power and rhizomatics, 52, 53
prejudice, 6, 17, 19–21, 23, 26–28, 41, 94, 140
preservice students, 169, 172, 174
proto-feminized whiteness, 68, 70, 75, 76, 78–82

race, 2–9, 17–29, 31–39, 41–47, 49–51, 54–56, 59–63, 67, 68, 70, 75, 77, 81, 82, 86, 88, 90, 94–97, 99, 114, 115, 117–119, 127–131, 133, 134, 139, 140, 142–144, 148, 150–153, 161, 167–171, 173–174, 176, 188, 192
  Australia, 2–9
  classification, 2, 3, 4, 24, 28, 29, 31
  and gender, 67, 68, 83, 85, 88, 94, 95, 97
  Great Chain of Being, 3
  hierarchies, 2, 3, 44, 59, 81, 88, 95, 133, 142
  and young children, 2, 6, 9, 17–30, 33, 37, 43, 45, 46, 68, 82, 83, 140, 143, 148
race-color, 4, 29, 46
race-power, 7, 28, 52–54, 61, 62, 86
  racial awareness, 20, 27, 38, 68, 84, 140, 151
racial bias, 19, 142
racial categorization, 32, 115

racial difference, 20, 26, 116, 124, 174
racial history Australia, 9–12
racial innocence, 18, 22, 42, 62, 82, 169
racial justice, 19
racial preferences, 22, 24–26, 39, 40, 67
racial segregation, 18
racial self-identification, 32, 39
racialization, 50–51, 60, 62, 83, 129, 134, 135
racing—definition, 2, 18
racism, 4, 5, 6, 8, 26, 27–29, 41, 42, 46, 47, 53, 62, 63, 82, 83, 84, 86, 88, 94, 96, 118, 130, 131–134, 137, 139–141, 144, 147–150, 152, 153, 158, 167–175, 177, 178, 189
  engaging with, 82, 120, 121, 124, 156, 160, 167, 172
  feeling comfortable, 167
research methods, 37, 38
  child interviews, 43, 46, 78
  ethnographic feedback, 31, 43, 45
  ethnographic research, 42, 43
  observation methods, 42–45
  PCCRCD project, 42–46, 51, 68, 69, 87–89
  self-portraiture, 42–44, 46, 47, 78, 188
  stories with dolls, 43, 46–47
rethinking identity, 32
rhizoanalysis, 50, 53, 60, 61, 64
  decalcomania, 51, 54
  lines of flight, 54, 55, 56, 58, 59, 62, 63, 64

self-identification, 32, 38, 39, 67
social identity theory, 27

theories of white identities, 36, 94

whiteness, 7, 25, 29, 34, 35, 37, 49, 50–63, 68, 70–72, 75, 76, 78, 79, 80, 85, 89, 92, 95, 96, 141, 155
  adults, 18, 52, 57, 58, 140
  Australian whiteness, 58
  children, 11, 19, 23–25, 28, 29, 39, 41, 53, 56–59
  contradictions, 119, 143, 151, 153
  early childhood educators, 53, 63
  mapping, 44, 51, 52, 61, 88, 114, 199
  normalization, 62, 124
  paradox, 117, 119
  power, 44, 51–53, 57, 59, 60
  white norm, 116, 124